Lecture Notes in Mathematics   1727

Editors:
A. Dold, Heidelberg
F. Takens, Groningen
B. Teissier, Paris

Springer
Berlin
Heidelberg
New York
Barcelona
Hong Kong
London
Milan
Paris
Singapore
Tokyo

Peter Kravanja   Marc Van Barel

Computing the Zeros
of Analytic Functions

Springer

Authors

Peter Kravanja
Marc Van Barel
Katholieke Universiteit Leuven
Department of Computer Science
Celestijnenlaan 200 A
3001 Heverlee, Belgium

E-mail:
Peter.Kravanja@na-net.ornl.gov
Marc.Van Barel@cs.kuleuven.ac.be

Cataloging-in-Publication Data applied for

Die Deutsche Bibliothek - CIP-Einheitsaufnahme

Kravanja, Peter:
Computing the zeros of analytic functions / Peter Kravanja ; Marc VanBarel. - Berlin ;
Heidelberg ; New York ; Barcelona ; Hong Kong ; London ; Milan ;
Paris ; Singapore ; Tokyo : Springer, 2000
(Lecture notes in mathematics ; 1727)
ISBN 3-540-67162-5

Mathematics Subject Classification (2000): Primary: 65H05 Secondary: 65E05,
65H10

ISSN 0075-8434
ISBN 3-540-67162-5 Springer-Verlag Berlin Heidelberg New York

Springer-Verlag is a company in the BertelsmannSpringer publishing group.
© Springer-Verlag Berlin Heidelberg 2000
Printed in Germany

The use of general descriptive names, registered names, trademarks, etc. in this
publication does not imply, even in the absence of a specific statement, that such
names are exempt from the relevant protective laws and regulations and therefore
free for general use.

Typesetting: Camera-ready T$_E$X output by the author
Printed on acid-free paper   SPIN: 10724923   41/3143/du   543210

Preface

In this book we consider the problem of computing zeros of analytic functions and several related problems in computational complex analysis.

We start by studying the problem of computing *all* the zeros of an analytic function f that lie inside a positively oriented Jordan curve γ. Our principal means of obtaining information about the location of the zeros is a certain symmetric bilinear form that can be evaluated via numerical integration along γ. This form involves the logarithmic derivative f'/f of f. Our approach could therefore be called a logarithmic residue based quadrature method. It can be seen as a continuation of the pioneering work done by Delves and Lyness. We shed new light on their approach by considering a different set of unknowns and by using the theory of formal orthogonal polynomials. Our algorithm computes not only approximations for the zeros but also their respective multiplicities. It does not require initial approximations for the zeros and we have found that it gives accurate results. The algorithm proceeds by solving generalized eigenvalue problems and a Vandermonde system. A Fortran 90 implementation is available (the package ZEAL). We also present an approach that uses only f and not its first derivative f'. These results are presented in Chapter 1.

In Chapter 2 we focus on the problem of locating clusters of zeros of analytic functions. We show how the approach presented in Chapter 1 can be used to compute approximations for the centre of a cluster and the total number of zeros in this cluster. We also attack the problem of computing all the zeros of f that lie inside γ in an entirely different way, based on rational interpolation at roots of unity. We show how the new approach complements the previous one and how it can be used effectively in case γ is the unit circle.

In Chapter 3 we show how our logarithmic residue based approach can be used to compute all the zeros and poles of a meromorphic function that lie in the interior of a Jordan curve.

In Chapter 4 we consider systems of analytic equations. A multidimensional logarithmic residue formula is available in the literature. This formula involves the integral of a differential form. We transform it into a sum of Riemann integrals and show how the zeros and their respective multiplicities can be computed from these integrals by solving a generalized eigenvalue problem that has Hankel structure and by solving several Vandermonde systems.

The problem of computing zeros of analytic functions leads to a rich blend of mathematics and numerical analysis. This book is a mixture of theoretical results (some of which are quite technical, for example, the differential forms that appear in the chapter on systems of analytic equations or the results that are based on rational interpolation), numerical analysis and algorithmic aspects, implementation heuristics, and polished software that is publicly available. We have found a lot of pleasure in researching and combining these elements and we hope that the reader will enjoy our work.

Peter Kravanja & Marc Van Barel

--

Katholieke Universiteit Leuven
Department of Computer Science
Celestijnenlaan 200 A, B-3001 Heverlee (Belgium)

`Peter.Kravanja@na-net.ornl.gov`
`Marc.VanBarel@cs.kuleuven.ac.be`

Contents

1. Zeros of analytic functions

In this chapter we will consider the problem of computing *all* the zeros of an analytic function f that lie in the interior of a Jordan curve γ. The algorithm that we will present computes not only approximations for the zeros but also their respective multiplicities. It doesn't require initial approximations for the zeros and gives accurate results. The algorithm is based on the theory of formal orthogonal polynomials. Its principal means of obtaining information about the location of the zeros is a certain symmetric bilinear form that can be evaluated via numerical integration along γ. This form involves the logarithmic derivative f'/f of f. Our approach could therefore be called a *logarithmic residue based quadrature method*. In the next chapters we will see how it can be used to locate clusters of zeros of analytic functions, to compute all the zeros and poles of a meromorphic function that lie in the interior of a Jordan curve, and to solve systems of analytic equations.

1.1 Introduction

Let W be a simply connected region in $\mathbb{C}$, $f : W \to \mathbb{C}$ analytic in W and γ a positively oriented Jordan curve in W that does not pass through any zero of f. We consider the problem of computing *all* the zeros of f that lie in the interior of γ, together with their respective multiplicities.

Our approach to this problem can be seen as a continuation of the pioneering work by Delves and Lyness [36]. Let N denote the total number of zeros of f that lie in the interior of γ, i.e., the number of zeros where each zero is counted according to its multiplicity. Suppose from now on that $N > 0$. Delves and Lyness considered the sequence $Z_1, \ldots, Z_N$ that consists of all the zeros of f that lie inside γ. Each zero is repeated according to its multiplicity. An easy calculation shows that the logarithmic derivative f'/f has a simple pole at each zero of f with residue equal to the multiplicity of the zero. Cauchy's Theorem implies that

$$N = \frac{1}{2\pi i} \int_\gamma \frac{f'(z)}{f(z)} \, dz. \tag{1.1}$$

This formula enables us to calculate N via numerical integration. Methods for the determination of zeros of analytic functions that are based on the numerical evaluation of integrals are called *quadrature methods*. A review of such methods was given by Ioakimidis [73]. Delves and Lyness considered the integrals

$$s_p := \frac{1}{2\pi i} \int_\gamma z^p \, \frac{f'(z)}{f(z)} \, dz, \qquad p = 0, 1, 2, \ldots .$$

The residue theorem implies that the s_p's are equal to the *Newton sums* of the unknown zeros,

$$s_p = Z_1^p + \cdots + Z_N^p, \qquad p = 0, 1, 2, \ldots . \tag{1.2}$$

In what follows we will assume that all the s_p's that are needed have been calculated. In particular, we will assume that the value of $N = s_0$ is known.

Delves and Lyness considered the monic polynomial of degree N that has zeros $Z_1, \ldots, Z_N$,

$$P_N(z) := \prod_{k=1}^N (z - Z_k) =: z^N + \sigma_1 \, z^{N-1} + \cdots + \sigma_N .$$

They called $P_N(z)$ the *associated polynomial* for the interior of γ. Its coefficients can be calculated via Newton's identities.

Theorem 1.1.1 (Newton's identities).

$$s_1 + \sigma_1 = 0$$
$$s_2 + s_1 \, \sigma_1 + 2 \, \sigma_2 = 0$$
$$\vdots$$
$$s_N + s_{N-1} \, \sigma_1 + \cdots + s_1 \, \sigma_{N-1} + N \, \sigma_N = 0.$$

Proof. An elegant proof was given by Carpentier and Dos Santos [31]. □

In this way they reduced the problem to the easier problem of computing the zeros of a polynomial. Unfortunately, the map from the Newton sums $s_1, \ldots, s_N$ to the coefficients $\sigma_1, \ldots, \sigma_N$ is usually ill-conditioned. Also, the polynomials that arise in practice may be such that small changes in the coefficients produce much larger changes in some of the zeros. This ill-conditioning of the map between the coefficients of a polynomial and its zeros was investigated by Wilkinson [124]. The location of the zeros determines their sensitivity to perturbations of the coefficients. Multiple zeros and very close zeros are extremely sensitive, but even a succession of moderately close zeros can result in severe ill-conditioning. Wilkinson states that ill-conditioning in polynomials cannot be overcome without, at some stage of the computation, resorting to high precision arithmetic.

If f has many zeros in the interior of γ, then the associated polynomial is of high degree and could be very ill-conditioned. Therefore, if N is large, one has to calculate the coefficients $\sigma_1, \ldots, \sigma_N$, and thus the integrals $s_1, \ldots, s_N$, very accurately. To avoid the use of high precision arithmetic and to reduce the number of integrand evaluations needed to approximate the s_p's, Delves and Lyness suggested to construct and solve the associated polynomial only if its degree is smaller than or equal to a preassigned number M. Otherwise, the interior of γ is subdivided or covered with a finite covering and the smaller regions are treated in turn. The choice of M involves a trade-off. If M is increased, then fewer regions have to be scanned. However, if M is chosen too large, then the resulting associated polynomial may be ill-conditioned. Delves and Lyness chose $M = 5$.

Botten, Craig and McPhedran [21] made a Fortran 77 implementation of the method of Delves and Lyness.

In some applications, the calculation of the derivative f' is more time-consuming than that of f. Delves and Lyness used an integration by parts to derive a formula for s_p that depends only on a multi-valued logarithm of f and not on f'. To apply this formula, they had to keep track of the sheet on which $\log f(z)$ lies as z runs along the curve γ. Unfortunately, in most cases it is impossible to do this in a completely reliable way, i.e., without accidentally overlooking any sheets. Carpentier and Dos Santos [31] and Davies [35] derived similar formulae. See also Ioakimidis and Anastasselou [76].

Instead of using Newton's identities to construct the associated polynomial, Li [93] considered (1.2) as a system of polynomial equations. He used a homotopy continuation method to solve this system.

What is wrong with these approaches, in our opinion, is that they consider the wrong set of unknowns. One should consider the mutually distinct zeros and their respective multiplicities *separately*. This is the approach that we will follow. Let n denote the number of mutually distinct zeros of f that lie inside γ. Let $z_1, \ldots, z_n$ be these zeros and $\nu_1, \ldots, \nu_n$ their respective multiplicities. The quadrature method that we will present generalizes the approach of Delves and Lyness. We will show how the mutually distinct zeros can be calculated by solving generalized eigenvalue problems. The value of n will be determined indirectly. Once n and $z_1, \ldots, z_n$ have been found, the problem becomes linear and the multiplicities $\nu_1, \ldots, \nu_n$ can be computed by solving a Vandermonde system.

The rest of this chapter is organized as follows. In the remainder of this section we will discuss how the total number of zeros can be calculated with certainty and we will give an overview of other approaches that were proposed for computing zeros of analytic functions. In Section 1.2 we will tackle our problem by using the theory of formal orthogonal polynomials. This section is devoted to theoretical considerations whereas our numerical algorithm will be presented in Section 1.3. We have found that this algorithm gives very accurate results. In Section 1.4 we will give numerical examples computed via a Matlab implementation whereas in Section 1.5 we will present a Fortran 90 implementation (the package ZEAL) and we will give more numerical examples. In Section 1.6 we will present a derivative-free approach that involves $1/f$ instead of f'/f and we will compare both approaches.

1.1.1 Computing the total number of zeros with certainty

The total number of zeros of f that lie inside γ is given by the integral

$$N = \frac{1}{2\pi i} \int_\gamma \frac{f'(z)}{f(z)}\, dz, \tag{1.3}$$

cf. Equation (1.1). By making the substitution $w := f(z)$ we obtain that

$$N = \frac{1}{2\pi i} \int_{f(\gamma)} \frac{1}{w}\, dw. \tag{1.4}$$

Here $f(\gamma)$ denotes the image of the curve γ under f. This is a closed curve that avoids the origin. The winding number of $f(\gamma)$ with respect to the origin is defined as the increase in the argument of $f(z)$ along γ divided by 2π,

$$n\big(f(\gamma), 0\big) := \frac{1}{2\pi} \big[\arg f(z)\big]_{z \in \gamma}.$$

Informally speaking, one can say that it is equal to the number of times that the curve $f(\gamma)$ "winds" itself around the origin. A classical theorem in complex analysis (see, e.g., Henrici [68, p. 233]) says that this winding number can be expressed as the integral that appears in the right-hand side of (1.4). Hence $N = n\big(f(\gamma), 0\big)$. This result is known as the "principle of the argument."

Example 1.1.1. Suppose that $f(z) = e^{3z} + 20z \cos z - 1$. Figure 1.1 shows the curve $f(\gamma)$ in case $\gamma = \{z \in \mathbb{C} : |z| = 2\}$. Clearly $N = 4$. If $\gamma = \{z \in \mathbb{C} : |z - 1| = 1/2\}$, then $N = 0$, as can be seen from Figure 1.2.   ◇

Earlier in this chapter we have suggested to calculate the total number of zeros N via numerical integration, i.e., by using (1.3). An alternative approach can be based on an algorithm for computing winding numbers. The range of the function arg is $(-\pi, \pi]$. If the increase in argument along the straight section

$$[\alpha, \beta] := \{z \in \mathbb{C} : z = t\alpha + (1-t)\beta, \quad 0 \le t \le 1\}, \qquad \alpha, \beta \in \mathbb{C},$$

satisfies

$$\big|\big[\arg f(z)\big]_{z \in [\alpha, \beta]}\big| \le \pi,$$

then

$$\big[\arg f(z)\big]_{z \in [\alpha, \beta]} = \arg\Big(\frac{f(\beta)}{f(\alpha)}\Big),$$

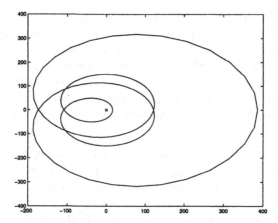

Fig. 1.1. The curve $f(\gamma)$ where $f(z) = e^{3z} + 20z \cos z - 1$ and $\gamma = \{z \in \mathbb{C} : |z| = 2\}$.

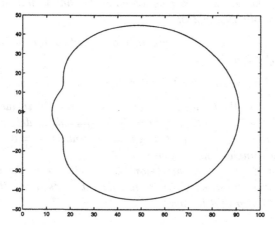

Fig. 1.2. The curve $f(\gamma)$ where $f(z) = e^{3z} + 20z \cos z - 1$ and $\gamma = \{z \in \mathbb{C} : |z - 1| = 1/2\}$.

as the reader may easily verify. Let us discretize the curve γ into the sequence of points $c_1, \ldots, c_G$. Define $c_{G+1} := c_1$. Then it follows that

$$N = \frac{1}{2\pi} \sum_{k=1}^{G} \arg\left(\frac{f(c_{k+1})}{f(c_k)}\right)$$

if

$$\left|\left[\arg f(z)\right]_{z \in [c_k, c_{k+1}]}\right| \leq \pi \tag{1.5}$$

for $k = 1, \ldots, G$. In other words, if condition (1.5) is satisfied, then N can be computed simply by evaluating f at the points $c_1, \ldots, c_G$. These considerations form the basis for Henrici's algorithm [68, pp. 239–241]. See also Ying and Katz [127].

Unfortunately, condition (1.5) may not be easy to verify for an arbitrary analytic function f. If the discretization of γ is inadequate, then the computed value of $n(f(\gamma), 0)$ and hence N may be wrong. In this sense Henrici's algorithm is unreliable and the same holds for numerical integration. Indeed, a finite number of functional or derivative values are not enough to determine the number of zeros of f, even if f is a polynomial. This was shown by Ying in his PhD thesis [126]. The next theorem is a slightly modified version of his result.

Theorem 1.1.2. *Let Ω be a simply connected region in $\mathbb{C}$. Let $l \geq 1$ be a positive integer and let $\zeta_1, \dots, \zeta_l \in \Omega$. Let $m_1, \dots, m_l \geq 0$ be nonnegative integers and suppose that numerical values are given for*

$$p(\zeta_1), p'(\zeta_1), \dots, p^{(m_1)}(\zeta_1), \ \dots, p(\zeta_l), p'(\zeta_l), \dots, p^{(m_l)}(\zeta_l) \qquad (1.6)$$

where p is only known to be a polynomial in Ω that is not identically zero in Ω. Then the following holds:

1. *The information in (1.6) is insufficient to determine the location of all the zeros of p that lie in Ω.*

   Let Γ be a Jordan curve in Ω that does not pass through any zero of p.

2. *Suppose that N, the total number of zeros of p that lie inside Γ, is known. Even then, the information in (1.6) is insufficient to determine the location of all the zeros of p that lie inside Γ, unless $N = 0$ or $N > 0$ and these zeros coincide with a subset of $\{\zeta_1, \dots, \zeta_l\}$.*

3. *If the points $\zeta_1, \dots, \zeta_l$ do not belong to the closure of the interior of Γ, then the information in (1.6) is insufficient to estimate the location of all the zeros of p that lie inside Γ.*

Proof.

1. Define $h(z)$ as $(z - \zeta_1)^{m_1 + 1} \cdots (z - \zeta_l)^{m_l + 1}$. Choose ζ_0 in $\Omega \setminus \{\zeta_1, \dots, \zeta_l\}$ such that $p(\zeta_0) \neq 0$ and let

   $$p_1(z) := p(z) - \frac{p(\zeta_0)}{h(\zeta_0)} h(z).$$

   Then p_1 is a polynomial in Ω and $p_1^{(k)}(\zeta_i) = p^{(k)}(\zeta_i)$ for $i = 1, \dots, l$ and $k = 0, 1, \dots, m_i$. However, $p_1(\zeta_0) = 0$ while $p(\zeta_0) \neq 0$.

2. Define

   $$\epsilon := \inf_{z \in \Gamma} |p(z)| \qquad \text{and} \qquad M := \sup_{z \in \Gamma} |h(z)|.$$

   Then $\epsilon > 0$ and $M > 0$. Let

   $$p_2(z) := p(z) + \frac{\epsilon}{2M} h(z).$$

Then p_2 is a polynomial in Ω and $p_2^{(k)}(\zeta_i) = p^{(k)}(\zeta_i)$ for $i = 1, \dots, l$ and $k = 0, 1, \dots, m_i$. Since $|p(z)| > |\frac{\varepsilon}{2M} h(z)|$ for all $z \in \Gamma$, Rouché's Theorem asserts that p_2 and p have the same total number of zeros in the interior of Γ. Let ζ be a zero of p that lies in the interior of Γ. If $\zeta \notin \{\zeta_1, \dots, \zeta_l\}$, then $p_2(\zeta) \neq 0$.

3. Define

$$\delta := \inf_{z \in \Gamma} |h(z)| \quad \text{and} \quad Q := \sup_{z \in \Gamma} |p(z)|.$$

Then $\delta > 0$ and $Q > 0$. Let

$$p_3(z) := p(z) + \frac{2Q}{\delta} h(z).$$

Then p_3 is a polynomial in Ω and $p_3^{(k)}(\zeta_i) = p^{(k)}(\zeta_i)$ for $i = 1, \dots, l$ and $k = 0, 1, \dots, m_i$. Since $|\frac{2Q}{\delta} h(z)| > |p(z)|$ for all $z \in \Gamma$, Rouché's Theorem asserts that p_3 and h have the same total number of zeros in the interior of Γ. Therefore, p_3 has no zeros in the interior of Γ.

This proves the theorem. $\square$

There exist several reliable approaches for computing N. However, they all assume that some kind of *global* information is available, which may not be the case in practice and hence these algorithms are in fact of little use.

- Ying and Katz [127] developed a reliable variant of Henrici's algorithm. They assume that an upper bound for $|f''(z)|$ along an arbitrary line segment is available.
- Herlocker and Ely [69] experimented with a numerical integration approach based on Simpson's rule and the corresponding formula for the integration error. This formula involves the fourth derivative of the integrand evaluated at an unknown point in the integration interval. Automatic differentiation combined with interval arithmetic enabled them to bound the integration error.
- The total number of zeros can also be computed as the topological degree of the mapping

$$F(x, y) := \big(\operatorname{Re} f(x + iy), \operatorname{Im} f(x + iy) \big)$$

with respect to the interior of γ (interpreted as a subset of $\mathbb{R}^2$) and the point $(0, 0)$. (We will not go into the details of degree theory. For an excellent introduction, we refer the interested reader to Lloyd's book [95].) Boult and Sikorski [22] considered the case that γ is the boundary of the unit square $[0, 1] \times [0, 1]$. They proved that, in case F satisfies the Lipschitz condition with constant $K > 0$ and if the infinity norm of F on γ is at least $d > 0$ where $K/(4d) \geq 1$, then at least $4\lfloor K/(4d) \rfloor$ function evaluations are needed to compute the topological degree. See also Traub, Wasilkowski and Woźniakowski [116, pp. 193–194].

As already mentioned, these reliable algorithms can seldom be used in practice since global information such as an upper bound for the modulus of a higher derivative of f or the Lipschitz constant of $F = (\text{Re } f, \text{Im } f)$ or a lower bound for the infinity norm of F on γ is usually not available. Therefore, although they may indeed give incorrect results, Henrici's algorithm and numerical integration are the only approaches that we can use to compute N. The output of Henrici's algorithm is always an integer but unfortunately one has no idea whether it's the correct integer or not. By using quadrature formulae with different degrees of accuracy, one can easily obtain an estimate of the quadrature error. Also, the size of the imaginary part of the computed approximation for N as well as the distance of the real part to the nearest integer give a clear indication of the error. (Of course, as the integral is known to be an integer, an approximation that has an error that is less than 0.5 is sufficient.) For these reasons, we prefer to use numerical integration instead of Henrici's algorithm.

1.1.2 An overview of other approaches

Let us conclude this section by giving a brief overview of other methods that were proposed for computing zeros of analytic functions.

Suppose that the zeros of f that lie inside γ are known to be simple. Then f can be written as

$$f(z) = \phi(z) \prod_{k=1}^{n} (z - z_k) \qquad (1.7)$$

where $\phi : W \to \mathbb{C}$ is analytic in W and does not vanish inside γ. Let $\alpha \in \mathbb{C}$ be an arbitrary point inside γ such that $f(\alpha) \neq 0$. Then in the interior of γ the function ϕ can be written as

$$\phi(z) = \exp \psi(z), \qquad z \in \text{int } \gamma, \qquad (1.8)$$

where ψ is defined by

$$\psi(z) := \frac{1}{2\pi i} \int_{\gamma} \frac{\log[(t - \alpha)^{-n} f(t)]}{t - z} \, dt, \qquad z \in \text{int } \gamma, \qquad (1.9)$$

cf. Smirnov [114]. The function ψ is analytic inside γ. Note that the logarithm in (1.9) is well defined as the winding number of $(z - \alpha)^{-n} f(z)$ with respect to γ and the origin is equal to zero.

Starting from (1.7) we find that

$$z_k = z - \frac{f(z)}{\phi(z) \prod_{j=1, j \neq k}^{n} (z - z_j)}$$

for $k = 1, \ldots, n$. Assume that distinct complex numbers $\zeta_1, \ldots, \zeta_n$ are reasonably good approximations for the zeros $z_1, \ldots, z_n$ of f. Putting $z = \zeta_k$ and substituting the zeros z_j by their approximations ζ_j $(j \neq k)$ we obtain

$$\hat{\zeta}_k := \zeta_k - \frac{f(\zeta_k)}{\phi(\zeta_k) \prod_{j=1, j \neq k}^{n} (\zeta_k - \zeta_j)} \tag{1.10}$$

for $k = 1, \ldots, n$. The point $\hat{\zeta}_k$ appears to be a new approximation for the zero z_k. Petković, Carstensen and Trajković [106] presented a simultaneous iterative method that is based on (1.10). They proved that the iteration converges quadratically if the initial approximations $\zeta_1^{(0)}, \ldots, \zeta_n^{(0)}$ are sufficiently close to the zeros $z_1, \ldots, z_n$ of f. The function ϕ is evaluated via numerical integration.

Since (1.7) and (1.8) imply that

$$\frac{f'(z)}{f(z)} = \psi'(z) + \sum_{k=1}^{n} \frac{1}{z - z_k}, \qquad z \notin \{z_1, \ldots, z_n\},$$

it follows that

$$z_k = z - \left[\frac{f'(z)}{f(z)} - \psi'(z) - \sum_{j=1, j \neq k}^{n} \frac{1}{z - z_j} \right]^{-1}, \qquad z \notin \{z_1, \ldots, z_n\},$$

for $k = 1, \ldots, n$. Putting again $z = \zeta_k$ and substituting the zeros z_j by their approximations ζ_j $(j \neq k)$ we obtain

$$\hat{\zeta}_k := \zeta_k - \left[\frac{f'(\zeta_k)}{f(\zeta_k)} - \psi'(\zeta_k) - \sum_{j=1, j \neq k}^{n} \frac{1}{\zeta_k - \zeta_j} \right]^{-1} \tag{1.11}$$

for $k = 1, \ldots, n$. Petković and Herceg [107] analysed the corresponding iterative method and proved that it has cubic convergence. They also presented a version of the algorithm that uses circular interval arithmetic (see also [105]). The derivative ψ' is given by

$$\psi'(z) = \frac{1}{2\pi i} \int_{\gamma} \frac{\log[(t - \alpha)^{-n} f(t)]}{(t - z)^2} \, dt, \qquad z \in \operatorname{int} \gamma,$$

and has to be evaluated via numerical integration.

Petković and Marjanović [108] generalized these results and gave simultaneous iterative methods that have order of convergence $p + 2$ $(p = 1, 2, \ldots)$ if p denotes the order of the highest derivative of f that appears in the iteration formula. These algorithms can be used only if the zeros of f are known to be simple and if sufficiently accurate initial approximations are available. Atanassova [12] considered the case of multiple zeros but assumed that the multiplicities are known in advance.

In case f is a polynomial, the iteration that corresponds to (1.10) is called the Durand-Kerner (or also Weierstrass) method whereas the iteration that corresponds to (1.11) is known as Aberth's method (cf. Bini [17]). The problem of calculating zeros of polynomials received a lot of interest in the past and is still a lively area of research. McNamee [97, 98] compiled an extensive bibliography on computing zeros of polynomials. Although polynomials are of course a special case of analytic functions, we will not give a comprehensive overview of all the methods that were proposed. Instead we refer the interested reader to the papers by Bini and Pan [19], Cardinal [29], Carstensen and Sakurai [32] and the survey paper by Pan [103] as well as the references cited therein. Recently, Bini, Gemignani and Meini [18] have presented an interesting method to compute a factor of a polynomial or of an analytic function that is given as a power series. Their approach is based on a matrix version of Koenig's Theorem and on cyclic reduction. It involves infinite Toeplitz matrices in block Hessenberg form.

In a number of papers and short notes, Anastasselou and Ioakimidis [3, 4, 5, 6, 7, 74, 75, 77, 78] considered the problem of computing zeros of sectionally analytic functions (i.e., functions that are analytic except for a finite number of discontinuity arcs). They proposed variations and generalizations of the method of Burniston and Siewert [27], which is based on the theory of Riemann-Hilbert boundary value problems (cf. Gakhov [47]). The authors focused on the function $\alpha + \beta z - z \tanh^{-1}(1/z)$ where $\alpha, \beta \in \mathbb{C}$ are parameters. This function appears in the theory of ferromagnetism. It has the discontinuity interval $[-1, 1]$. Already for this example, the approach of Anastasselou and Ioakimidis requires a lot of specific analytical calculations. Therefore, it is unclear how their method could lead to a 'black box' algorithm that can handle arbitrary functions. Also, although multiple zeros are not a problem, their approach cannot calculate multiplicities.

Yakoubsohn [125] proposed an exclusion method for computing zeros of analytic functions. Unfortunately, his exclusion function is difficult to evaluate and multiple zeros require special treatment. He applied his algorithm only to polynomials. See also Ying and Katz [128].

1.2 Formal orthogonal polynomials

Let $\mathcal{P}$ be the linear space of polynomials with complex coefficients. We define a symmetric bilinear form

$$\langle \cdot, \cdot \rangle : \mathcal{P} \times \mathcal{P} \to \mathbb{C}$$

by setting

$$\langle \phi, \psi \rangle := \frac{1}{2\pi i} \int_\gamma \phi(z)\psi(z)\frac{f'(z)}{f(z)}\,dz = \sum_{k=1}^n \nu_k \phi(z_k)\psi(z_k) \qquad (1.12)$$

for any two polynomials $\phi, \psi \in \mathcal{P}$. The latter equality follows from the fact that f'/f has a simple pole at z_k with residue ν_k for $k = 1, \ldots, n$. Note that $\langle \cdot, \cdot \rangle$ can be evaluated via numerical integration along γ. In what follows, we will assume that all the "inner products" $\langle \phi, \psi \rangle$ that are needed have been calculated. Let $s_p := \langle 1, z^p \rangle$ for $p = 0, 1, 2, \ldots$. These ordinary moments are equal to the *Newton sums* of the unknown zeros,

$$s_p = \sum_{k=1}^{n} \nu_k z_k^p, \qquad p = 0, 1, 2, \ldots.$$

In particular, $s_0 = \nu_1 + \cdots + \nu_n = N$, the total number of zeros. Hence, we may assume that the value of N is known. Let H_k be the $k \times k$ Hankel matrix

$$H_k := \left[s_{p+q} \right]_{p,q=0}^{k-1} = \begin{bmatrix} s_0 & s_1 & \cdots & s_{k-1} \\ s_1 & & \cdots & \vdots \\ \vdots & \cdots & & \vdots \\ s_{k-1} & \cdots & \cdots & s_{2k-2} \end{bmatrix}$$

for $k = 1, 2, \ldots$. Let H be the infinite Hankel matrix

$$H := \left[s_{p+q} \right]_{p,q \geq 0}.$$

Observe that the form $\langle \cdot, \cdot \rangle$ is completely determined by the sequence of moments $(s_p)_{p \geq 0}$.

A monic polynomial φ_t of degree $t \geq 0$ that satisfies

$$\langle z^k, \varphi_t(z) \rangle = 0, \qquad k = 0, 1, \ldots, t-1, \tag{1.13}$$

is called a *formal orthogonal polynomial* (FOP) of degree t. (Observe that condition (1.13) is void for $t = 0$.) The adjective *formal* emphasizes the fact that, in general, the form $\langle \cdot, \cdot \rangle$ does not define a true inner product. An important consequence of this fact is that, in contrast to polynomials that are orthogonal with respect to a true inner product, formal orthogonal polynomials need not exist or need not be unique for every degree. (For details, see for example Draux [37, 38], Gutknecht [66, 67] or Gragg and Gutknecht [61].) If (1.13) is satisfied and φ_t is unique, then φ_t is called a *regular* FOP and t a *regular index*. If we set

$$\varphi_t(z) =: u_{0,t} + u_{1,t} z + \cdots + u_{t-1,t} z^{t-1} + z^t$$

then condition (1.13) translates into the Yule-Walker system

$$\begin{bmatrix} s_0 & s_1 & \cdots & s_{t-1} \\ s_1 & & \cdots & \vdots \\ \vdots & \cdots & & \vdots \\ s_{t-1} & \cdots & \cdots & s_{2t-2} \end{bmatrix} \begin{bmatrix} u_{0,t} \\ u_{1,t} \\ \vdots \\ u_{t-1,t} \end{bmatrix} = - \begin{bmatrix} s_t \\ s_{t+1} \\ \vdots \\ s_{2t-1} \end{bmatrix}. \tag{1.14}$$

Hence, the regular FOP of degree $t \geq 1$ exists if and only if the matrix H_t is nonsingular. Thus, the rank profile of H determines which regular FOPs exist. If t is a regular index, then

$$\varphi_t(z) = \frac{1}{\det H_t} \begin{vmatrix} s_0 & s_1 & \cdots & s_{t-1} & 1 \\ s_1 & & \cdots & \vdots & z \\ \vdots & \cdots & & \vdots & \vdots \\ s_{t-1} & \cdots & \cdots & s_{2t-2} & z^{t-1} \\ s_t & \cdots & \cdots & s_{2t-1} & z^t \end{vmatrix},$$

as one can easily verify. Note that this determinant expression implies that

$$\langle \varphi_t, \varphi_t \rangle = \frac{\det H_{t+1}}{\det H_t}. \tag{1.15}$$

The following theorem characterizes n, the number of mutually distinct zeros. It enables us, theoretically at least, to calculate n as rank H_N.

Theorem 1.2.1. $n = \operatorname{rank} H_{n+p}$ *for every nonnegative integer p. In particular, $n = \operatorname{rank} H_N$.*

Proof. Let p be a nonnegative integer. The matrix H_{n+p} can be written as

$$H_{n+p} = \sum_{k=1}^{n} \nu_k \begin{bmatrix} 1 & \cdots & z_k^{n+p-1} \\ \vdots & & \vdots \\ z_k^{n+p-1} & \cdots & z_k^{2(n+p)-2} \end{bmatrix}$$

$$= \sum_{k=1}^{n} \nu_k \begin{bmatrix} 1 \\ \vdots \\ z_k^{n+p-1} \end{bmatrix} \begin{bmatrix} 1 & \cdots & z_k^{n+p-1} \end{bmatrix}.$$

This implies that rank $H_{n+p} \leq n$. However, H_n is nonsingular. Indeed, as one can easily verify, the matrix H_n can be factorized as $H_n = V_n D_n V_n^T$ where V_n is the Vandermonde matrix $V_n := [z_s^r]_{r=0,s=1}^{n-1,n}$ and D_n is the diagonal matrix $D_n := \operatorname{diag}(\nu_1, \ldots, \nu_n)$. Therefore rank $H_{n+p} \geq n$. It follows that rank $H_{n+p} = n$. $\qquad\square$

Therefore H_n is nonsingular whereas H_t is singular for $t > n$. Note that $H_1 = [s_0]$ is nonsingular by assumption. The regular FOP of degree 1 exists and is given by $\varphi_1(z) = z - \mu$ where

$$\mu := \frac{s_1}{s_0} = \frac{\sum_{k=1}^{n} \nu_k z_k}{\sum_{k=1}^{n} \nu_k}$$

is the arithmetic mean of the zeros. Theorem 1.2.1 implies that the regular FOP φ_n of degree n exists and tells us also that regular FOPs of degree larger than n do not exist. The polynomial φ_n is easily seen to be

$$\varphi_n(z) = (z - z_1) \cdots (z - z_n). \tag{1.16}$$

It is the monic polynomial of degree n that has $z_1, \ldots, z_n$ as simple zeros. Its coefficients can be calculated by solving an $n \times n$ Yule-Walker system. This polynomial has the peculiar property that it is orthogonal to *all* polynomials (including itself),

$$\langle z^k, \varphi_n(z) \rangle = 0, \qquad k = 0, 1, 2, \ldots . \tag{1.17}$$

Note 1.2.1. Kronecker's Theorem [104, p.37] tells us that the infinite Hankel matrix H has finite rank if and only if its *symbol*, which is defined as the formal Laurent series

$$\frac{s_0}{z} + \frac{s_1}{z^2} + \frac{s_2}{z^3} + \cdots ,$$

represents a rational function of z. This is indeed the case. It is easily seen that

$$\sum_{k=1}^n \frac{\nu_k}{z - z_k} = \frac{s_0}{z} + \frac{s_1}{z^2} + \frac{s_2}{z^3} + \cdots \qquad \text{near } z = \infty. \tag{1.18}$$

In system theory the problem of reconstructing the rational function that appears in the left-hand side of (1.18) from the sequence of moments $(s_p)_{p \geq 0}$ is called a minimal realization problem. In that context the moments are called Markov parameters. For more details and the connection with continued fractions, Padé approximation, the Euclidean algorithm for formal Laurent series and the Berlekamp-Massey algorithm, see for example [24, 25, 26, 37, 61, 81]. Note that the left-hand side of (1.18) is a rational function of type $[n - 1/n]$ and that its denominator polynomial is given by $\varphi_n(z)$.

Once n is known, the mutually distinct zeros $z_1, \ldots, z_n$ can be calculated by solving a generalized eigenvalue problem. Indeed, let $H_n^<$ be the Hankel matrix

$$H_n^< := \begin{bmatrix} s_1 & s_2 & \cdots & s_n \\ s_2 & & \ddots & \vdots \\ \vdots & \ddots & \ddots & \vdots \\ s_n & \cdots & \cdots & s_{2n-1} \end{bmatrix}.$$

Theorem 1.2.2. *The eigenvalues of the pencil $H_n^< - \lambda H_n$ are given by $z_1, \ldots, z_n$.*

Proof. Suppose that $\varphi_n(z) =: u_{0,n} + u_{1,n}z + \cdots + u_{n-1,n}z^{n-1} + z^n$. Then (1.16) implies that the zeros $z_1, \ldots, z_n$ are given by the eigenvalues of the companion matrix

$$C_n := \begin{bmatrix} 0 & 0 & \cdots & 0 & -u_{0,n} \\ 1 & 0 & \cdots & 0 & -u_{1,n} \\ 0 & 1 & \ddots & \vdots & \vdots \\ \vdots & & \ddots & 0 & -u_{n-2,n} \\ 0 & \cdots & 0 & 1 & -u_{n-1,n} \end{bmatrix}.$$

Let $\lambda^\star$ be an eigenvalue of C_n and x a corresponding eigenvector. As H_n is regular, we may conclude that

$$C_n x = \lambda^\star x \Leftrightarrow H_n C_n x = \lambda^\star H_n x.$$

Using (1.14) one can easily verify that $H_n C_n = H_n^<$. This proves the theorem.

Another proof goes as follows. As in the proof of Theorem 1.2.1, let V_n be the Vandermonde matrix

$$V_n := \begin{bmatrix} 1 & \cdots & 1 \\ z_1 & \cdots & z_n \\ \vdots & & \vdots \\ z_1^{n-1} & \cdots & z_n^{n-1} \end{bmatrix}$$

and let $D_n := \operatorname{diag}(\nu_1, \ldots, \nu_n)$. Also, let $D_n^{(1)} := \operatorname{diag}(\nu_1 z_1, \ldots, \nu_n z_n)$. Then the matrices H_n and $H_n^<$ can be factorized as

$$H_n = V_n D_n V_n^T \quad \text{and} \quad H_n^< = V_n D_n^{(1)} V_n^T.$$

Let $\lambda^\star$ be an eigenvalue of the pencil $H_n^< - \lambda H_n$ and x a corresponding eigenvector. Then

$$\begin{aligned} & H_n^< x = \lambda^\star H_n x \\ \Leftrightarrow\ & V_n D_n^{(1)} V_n^T x = \lambda^\star V_n D_n V_n^T x \\ \Leftrightarrow\ & D_n^{(1)} y = \lambda^\star D_n y \quad \text{if } y := V_n^T x \\ \Leftrightarrow\ & \operatorname{diag}(z_1, \ldots, z_n) y = \lambda^\star y. \end{aligned}$$

This proves the theorem.   □

Note 1.2.2. Recently, Golub, Milanfar and Varah [60] have presented a stable numerical solution to the problem of reconstructing a polygonal shape from moments. This problem has many applications including tomographic reconstruction and geophysical inversion. In the latter application, it is of interest to reconstruct the shape and (possibly) density of a gravitational anomaly from discrete measurements of the exterior gravitational field at spatially separated points. The authors' approach is based on the matrix pencil that appears in Theorem 1.2.2. To solve the generalized eigenvalue problem, they have developed an efficient variant of the QZ algorithm that takes into account the way the matrices H_n and $H_n^<$ are related.

Once $z_1, \ldots, z_n$ have been found, the multiplicities $\nu_1, \ldots, \nu_n$ can be computed by solving the Vandermonde system

$$\begin{bmatrix} 1 & \cdots & 1 \\ z_1 & \cdots & z_n \\ \vdots & & \vdots \\ z_1^{n-1} & \cdots & z_n^{n-1} \end{bmatrix} \begin{bmatrix} \nu_1 \\ \nu_2 \\ \vdots \\ \nu_n \end{bmatrix} = \begin{bmatrix} s_0 \\ s_1 \\ \vdots \\ s_{n-1} \end{bmatrix}. \tag{1.19}$$

This can be done via the algorithm of Gohberg and Koltracht [59]. This algorithm takes full account of the structure of a Vandermonde matrix and is not only faster but also more accurate than general purpose algorithms such as Gaussian elimination with partial pivoting. It has arithmetic complexity $\mathcal{O}(n^2)$.

Note 1.2.3. Vandermonde matrices are often very ill-conditioned. Gautschi wrote many papers on this subject, see for example [33, 48, 49, 51, 52, 55, 56]. In our case, however, the components of the solution vector of (1.19) are known to be integers, and therefore there is no problem, even if the linear system (1.19) happens to be ill-conditioned, as long as the computed approximations for the components of the solution vector have an absolute error that is less than 0.5. We remark that Hankel matrices also have the reputation of being ill-conditioned [117].

Theorem 1.2.1 and 1.2.2 suggest the following approach to compute n and $z_1, \ldots, z_n$. Start by computing the total number of zeros N. Next, compute $s_1, \ldots, s_{2N-2}$. As already mentioned, this can be done via numerical integration along γ. The number of mutually distinct zeros is then calculated as the rank of H_N, $n = \text{rank} \, H_N$. Finally, the zeros $z_1, \ldots, z_n$ are obtained by solving a generalized eigenvalue problem. Unfortunately, this approach has several disadvantages:

- Theoretically the $N-n$ smallest singular values of H_N are equal to zero. In practice, this will not be the case, and it may be difficult to determine the rank of H_N and hence the value of n in case the gap between the computed approximations for the zero singular values and the nonzero singular values is too small.
- The approximations for $z_1, \ldots, z_n$ obtained via Theorem 1.2.2 may not be very accurate. Indeed, the mapping from the Newton sums to the zeros and their respective multiplicities,

$$(s_0, s_1, \ldots, s_{2n-1}) \mapsto (z_1, \ldots, z_n, \nu_1, \ldots, \nu_n), \qquad (1.20)$$

is usually very ill-conditioned. (See for example the papers by Gautschi [50, 53, 54] who studied the conditioning of (1.20) in the context of Gauss quadrature formulae. For a recent paper on this subject, we refer to Beckermann and Bourreau [13].) Indeed, a classical adage in numerical analysis says that one should avoid the use of ordinary moments.

In Section 1.3 we will present an algorithm that gives more accurate approximations for $z_1, \ldots, z_n$. The idea is the following. The inner products that appear in the Hankel matrices H_n and $H_n^<$ are related to the standard monomial basis. Why not consider a different basis? In other words, let us try to use modified moments instead of ordinary moments. The fact that

$$H_n = \left[\langle z^p, z^q \rangle \right]_{p,q=0}^{n-1} \quad \text{and} \quad H_n^< = \left[\langle z^p, zz^q \rangle \right]_{p,q=0}^{n-1}$$

suggests that we should consider the matrices

$$[\langle\psi_p,\psi_q\rangle]_{p,q=0}^{n-1} \quad \text{and} \quad [\langle\psi_p,\psi_1\psi_q\rangle]_{p,q=0}^{n-1} \qquad (1.21)$$

where ψ_k is a polynomial of degree k for $k = 0, 1, \ldots, n - 1$. Of course, even if we succeed in writing Theorem 1.2.2 in terms of (1.21), the question remains which polynomials ψ_k to choose. We have found that very accurate results are obtained if we use the formal orthogonal polynomials. In other words, the zeros of $\dot\varphi_n(z)$ will be computed from inner products that involve $\varphi_0(z), \varphi_1(z), \ldots, \varphi_{n-1}(z)$. The value of n will be determined indirectly.

All this will be explained in more detail in Section 1.3. Let us conclude this section by discussing the orthogonality properties of FOPs. This will enable us to define the matrices (1.21) and to examine their structure.

If H_n is strongly nonsingular, i.e., if all its leading principal submatrices are nonsingular, then we have a full set $\{\varphi_0, \varphi_1, \ldots, \varphi_n\}$ of regular FOPs.

What happens if H_n is not strongly nonsingular and thus there is no full set of regular FOPs? Let $\{k_j\}_{j=0}^{J}$ be the set of all regular indices, with

$$k_0 < k_1 < \cdots < k_J.$$

Then $k_0 = 0$, $k_1 = 1$ and $k_J = n$. By filling up the gaps in the sequence of existing regular FOPs it is possible to define a sequence $\{\varphi_t\}_{t=0}^{\infty}$, with φ_t a monic polynomial of degree t, such that if these polynomials are grouped into blocks according to the sequence of regular indices, then polynomials belonging to different blocks are orthogonal with respect to (1.12). More precisely, define $\{\varphi_t\}_{t=0}^{\infty}$ as follows. If t is a regular index, then let φ_t be the regular FOP of degree t. Else define φ_t as $\varphi_r\psi_{t,r}$ where r is the largest regular index less than t and $\psi_{t,r}$ is an arbitrary monic polynomial of degree $t - r$. In the latter case φ_t is called an *inner polynomial*. If $\psi_{t,r}(z) = z^{t-r}$ then we say that φ_t is defined *by using the standard monomial basis*. These polynomials $\{\varphi_t\}_{t=0}^{\infty}$ can be grouped into $J + 1$ blocks

$$\Phi^{(0)} := [\varphi_0]$$
$$\Phi^{(1)} := [\varphi_1 \; \varphi_2 \; \cdots \; \varphi_{k_2-1}]$$
$$\Phi^{(2)} := [\varphi_{k_2} \; \varphi_{k_2+1} \; \cdots \; \varphi_{k_3-1}]$$
$$\vdots \qquad \vdots$$
$$\Phi^{(J-1)} := [\varphi_{k_{J-1}} \; \varphi_{k_{J-1}+1} \; \cdots \; \varphi_{k_J-1}]$$
$$\Phi^{(J)} := [\varphi_n \; \varphi_{n+1} \; \cdots \;].$$

Note that each block starts with a regular FOP and that the remaining polynomials in each block are inner polynomials. The pth block has length $l_p := k_{p+1} - k_p$ for $p = 0, 1, \ldots, J - 1$. Let

$$\langle \Psi, \Phi \rangle := \begin{bmatrix} \langle \psi_0, \phi_0 \rangle & \cdots & \langle \psi_0, \phi_q \rangle \\ \vdots & & \vdots \\ \langle \psi_p, \phi_0 \rangle & \cdots & \langle \psi_p, \phi_q \rangle \end{bmatrix} \in \mathbb{C}^{(p+1) \times (q+1)}$$

for any two row vectors

$$\Psi := [\psi_0 \ \psi_1 \ \cdots \ \psi_p] \qquad \text{and} \qquad \Phi := [\phi_0 \ \phi_1 \ \cdots \ \phi_q]$$

of polynomials in $\mathcal{P}$.

Theorem 1.2.3. *The following block orthogonality relations hold:*

$$\langle \Phi^{(p)}, \Phi^{(q)} \rangle = \begin{cases} 0_{l_p \times l_q} & \text{if } p \neq q \\ \delta_p & \text{if } p = q \end{cases} \quad \text{for } p, q = 0, 1, \ldots, J-1.$$

The matrix $\delta_p \in \mathbb{C}^{l_p \times l_p}$ is nonsingular and symmetric for $p = 0, 1, \ldots, J - 1$. Its entries are equal to zero above the main antidiagonal and equal to $\langle z^{k_p + l_p - 1}, \varphi_{k_p} \rangle$ along the main antidiagonal. Also, if all the inner polynomials of the block $\Phi^{(p)}$ where $p \in \{1, \ldots, J-1\}$ are defined by using the standard monomial basis, then δ_p is a Hankel matrix.

Proof. This result is well-known in the theory of FOPs. However, since readers who are less familiar with formal orthogonal polynomials than with the literature concerning the computation of zeros may find themselves slightly overwhelmed by the amount of information given in, for example, the survey papers by Gutknecht [66, 67] or the book by Bultheel and Van Barel [26], we prefer to include a proof.

The proof is by induction. Obviously, $\langle \Phi^{(0)}, \Phi^{(0)} \rangle = [\langle 1, 1 \rangle] = [s_0]$ is nonsingular. Now suppose that the theorem holds for $p, q = 0, 1, \ldots, k - 1$ where $k \in \{1, \ldots, J-1\}$. Consider the block $\Phi^{(k)}$. Let us call the first polynomial of this block φ_r and let l be the length of this block,

$$\Phi^{(k)} = [\varphi_r \ \varphi_{r+1} \ \cdots \ \varphi_{r+l-1}].$$

Then the matrices $H_{r+1}, \ldots, H_{r+l-1}$ are singular while H_r and H_{r+l} are nonsingular. By symmetry considerations it suffices to prove that

$$\hat{K} := \begin{bmatrix} \langle \varphi_0, \varphi_r \rangle & \cdots & \langle \varphi_0, \varphi_{r+l-1} \rangle \\ \langle \varphi_1, \varphi_r \rangle & \cdots & \langle \varphi_1, \varphi_{r+l-1} \rangle \\ \vdots & & \vdots \\ \langle \varphi_{r-1}, \varphi_r \rangle & \cdots & \langle \varphi_{r-1}, \varphi_{r+l-1} \rangle \end{bmatrix} = 0_{r \times l}$$

and that the matrix

$$\hat{\delta} := \begin{bmatrix} \langle \varphi_r, \varphi_r \rangle & \cdots & \langle \varphi_r, \varphi_{r+l-1} \rangle \\ \vdots & & \vdots \\ \langle \varphi_{r+l-1}, \varphi_r \rangle & \cdots & \langle \varphi_{r+l-1}, \varphi_{r+l-1} \rangle \end{bmatrix}$$

is nonsingular and has the other properties mentioned. Let F_l be the $l \times l$ unit upper triangular matrix that contains the coefficients of the polynomials

$$1, \psi_{r+1,r}, \ldots, \psi_{r+l-1,r}$$

(used in the definition of the inner polynomials of the block $\Phi^{(k)}$). Then

$$\hat{K} = K F_l \qquad \text{and} \qquad \hat{\delta} = F_l^T \delta F_l$$

if the matrices K and δ are defined as

$$K := \begin{bmatrix} \langle \varphi_0, \varphi_r \rangle & \langle \varphi_0, z\varphi_r \rangle & \cdots & \langle \varphi_0, z^{l-1}\varphi_r \rangle \\ \vdots & \vdots & & \vdots \\ \langle \varphi_{r-1}, \varphi_r \rangle & \langle \varphi_{r-1}, z\varphi_r \rangle & \cdots & \langle \varphi_{r-1}, z^{l-1}\varphi_r \rangle \end{bmatrix}$$

and

$$\delta := \begin{bmatrix} \langle \varphi_r, \varphi_r \rangle & \langle \varphi_r, z\varphi_r \rangle & \cdots & \langle \varphi_r, z^{l-1}\varphi_r \rangle \\ \langle z\varphi_r, \varphi_r \rangle & & \cdot^{\cdot^{\cdot}} & \vdots \\ \vdots & \cdot^{\cdot^{\cdot}} & & \vdots \\ \langle z^{l-1}\varphi_r, \varphi_r \rangle & \cdots & \cdots & \langle z^{l-1}\varphi_r, z^{l-1}\varphi_r \rangle \end{bmatrix}.$$

(In other words, K and δ correspond to the situation where all the inner polynomials in our block are defined by using the standard monomial basis.) Therefore we will first study the matrices K and δ. Observe that δ is a Hankel matrix. As φ_r is a regular FOP, we may conclude that $\langle \varphi_s, z^t \varphi_r \rangle = 0$ for $t \geq 0$ and $s = 0, 1, \ldots, r - t - 1$. Thus

$$K = \begin{bmatrix} 0 & \cdots & \cdots & 0 \\ \vdots & \ddots & & \vdots \\ \vdots & & \ddots & \vdots \\ 0 & \cdots & \cdots & 0 \\ \vdots & & \cdot^{\cdot^{\cdot}} & \times \\ \vdots & \cdot^{\cdot^{\cdot}} & \cdot^{\cdot^{\cdot}} & \vdots \\ 0 & \times & \cdots & \times \end{bmatrix}. \tag{1.22}$$

Let us consider the first antidiagonal of K whose entries we have not yet proven to be equal to zero. As φ_r is orthogonal to all polynomials of degree $\leq r - 1$, all these entries are equal. Indeed,

$$\langle \varphi_{r-1}, z\varphi_r \rangle = \langle \varphi_{r-2}, z^2\varphi_r \rangle = \cdots = \langle \varphi_{r-l+1}, z^{l-1}\varphi_r \rangle = \langle z^r, \varphi_r \rangle.$$

Note that $\langle z^r, \varphi_r \rangle = \langle \varphi_r, \varphi_r \rangle$, the entry in the upper left corner of δ. As H_{r+1} is singular and $\langle \varphi_r, \varphi_r \rangle = \det H_{r+1} / \det H_r$, it follows that $\langle \varphi_r, \varphi_r \rangle = 0$. This implies that φ_r is in fact orthogonal to all polynomials of degree $\leq r$ and that the upper left entry of δ as well as all the entries on our antidiagonal of K are equal to zero. Now we continue with the next antidiagonal of K. The fact that $\langle \varphi_r, \varphi_r \rangle = 0$ implies that all its entries are equal to $\langle z^{r+1}, \varphi_r \rangle = \langle z\varphi_r, \varphi_r \rangle$. One can easily check that

$$R_{r+2}^T H_{r+2} R_{r+2} = \left[\langle \varphi_p, \varphi_q \rangle \right]_{p,q=0}^{r+1}$$

$$= \operatorname{diag}(\delta_0, \delta_1, \ldots, \delta_{k-1}) \oplus F_2^T \begin{bmatrix} 0 & \langle z^{r+1}, \varphi_r \rangle \\ \langle z^{r+1}, \varphi_r \rangle & \langle z\varphi_r, z\varphi_r \rangle \end{bmatrix} F_2$$

if R_{r+2} is the unit upper triangular matrix that contains the coefficients of the polynomials $\varphi_0, \varphi_1, \ldots, \varphi_{r+1}$. As H_{r+2} is singular and $\delta_0, \delta_1, \ldots, \delta_{k-1}$ are nonsingular, it follows that $\langle z^{r+1}, \varphi_r \rangle = 0$. Thus φ_r is orthogonal to all polynomials of degree $\leq r+1$ and all the entries on our antidiagonal of K as well as two additional entries of δ are equal to zero. And so on. Eventually we will find that all the entries of K that are marked $\times$ in (1.22) are determined by the first $l-1$ entries of the first row of δ "in a Hankel way," i.e., by shifting these entries to the north-east. We will also find that $\langle z^{r+2}, \varphi_r \rangle = \cdots = \langle z^{r+l-2}, \varphi_r \rangle = 0$, i.e., φ_r is orthogonal to all polynomials of degree $\leq r+l-2$ (because $H_{r+3}, \ldots, H_{r+l-1}$ are singular) while $\langle z^{r+l-1}, \varphi_r \rangle \neq 0$ (because H_{r+l} is nonsingular). Therefore $\hat{K} = K = 0_{r \times l}$ and δ is a nonsingular lower triangular Hankel matrix. One can easily verify that this implies that $\hat{\delta}$ is indeed nonsingular, symmetric, zero above the main antidiagonal, and equal to $\langle z^{r+l-1}, \varphi_r \rangle$ along the main antidiagonal. This proves the theorem. □

Note that $\langle \Phi^{(p)}, \Phi^{(J)} \rangle = 0_{l_p \times \infty}$ for $p = 0, 1, \ldots, J$ if we set $l_J := \infty$.

Let G be the infinite Gram matrix

$$G := \left[\langle \varphi_p, \varphi_q \rangle \right]_{p,q \geq 0}$$

and let G_k be its $k \times k$ leading principal submatrix for $k = 1, 2, \ldots$. Then

$$G = G_n \oplus 0_{\infty \times \infty}.$$

The matrix G_n is nonsingular and block diagonal, $G_n = \operatorname{diag}(\delta_0, \delta_1, \ldots, \delta_{J-1})$. Let $G^{(1)}$ be the infinite matrix

$$G^{(1)} := \left[\langle \varphi_p, \varphi_1 \varphi_q \rangle \right]_{p,q \geq 0}$$

and let $G_k^{(1)}$ be its $k \times k$ leading principal submatrix for $k = 1, 2, \ldots$. Then

$$G^{(1)} = G_n^{(1)} \oplus 0_{\infty \times \infty}.$$

Note that G and $G^{(1)}$ are both symmetric. As already mentioned, the matrices G_n and $G_n^{(1)}$ will play an important role in Section 1.3. The following theorem examines the structure of $G_n^{(1)}$. Let us agree to call a matrix $A = [a_{p,q}]_{p,q=0}^{l-1} \in \mathbb{C}^{l \times l}$ *lower anti-Hessenberg* if $a_{p,q} = 0$ whenever $p + q < l - 2$, i.e., a matrix will be called lower anti-Hessenberg if its entries are equal to zero along all the antidiagonals that lie above the main antidiagonal, except for the antidiagonal that precedes the main antidiagonal.

Theorem 1.2.4. *The following block orthogonality relations hold:*

$$\langle \Phi^{(p)}, \varphi_1 \Phi^{(q)} \rangle = \begin{cases} 0_{l_p \times l_q} & \text{if } |p-q| > 1 \\ \kappa_p & \text{if } p = q+1 \\ \kappa_q^T & \text{if } p = q-1 \\ \delta_p^{(1)} & \text{if } p = q \end{cases} \quad \text{for } p, q = 0, 1, \ldots, J-1.$$

In other words, the matrix $G_n^{(1)}$ is block tridiagonal,

$$G_n^{(1)} = \begin{bmatrix} \delta_0^{(1)} & \kappa_1^T & & & \\ \kappa_1 & \delta_1^{(1)} & \kappa_2^T & & \\ & \ddots & \ddots & \ddots & \\ & & \kappa_{J-2} & \delta_{J-2}^{(1)} & \kappa_{J-1}^T \\ & & & \kappa_{J-1} & \delta_{J-1}^{(1)} \end{bmatrix}.$$

The matrix $\delta_p^{(1)} \in \mathbb{C}^{l_p \times l_p}$ is symmetric and lower anti-Hessenberg for $p = 0, 1, \ldots, J-1$. Its entries are equal to $\langle z^{k_p + l_p - 1}, \varphi_{k_p} \rangle$ along its first nonzero antidiagonal. Also, if all the inner polynomials of the block $\Phi^{(p)}$ where $p \in \{1, \ldots, J-1\}$ are defined by using the standard monomial basis, then $\delta_p^{(1)}$ is a Hankel matrix. Note that $\delta_0^{(1)} = [0]$. All the entries of the matrix κ_p where $p \in \{1, \ldots, J-1\}$ are equal to zero, except for the entry in the south-east corner, which is equal to $\langle z^{k_p + l_p - 1}, \varphi_{k_p} \rangle$.

Proof. The proof is left to the reader. Use the results obtained in the proof of Theorem 1.2.3. □

Thus, for example, the matrices G_n and $G_n^{(1)}$ may look like

$$G_n = \begin{bmatrix} \otimes & & & & & & & & & \\ & 0 & 0 & \otimes & & & & & & \\ & 0 & \otimes & \times & & & & & & \\ & \otimes & \times & \times & & & & & & \\ & & & & 0 & 0 & \otimes & & & \\ & & & & 0 & \otimes & \times & & & \\ & & & & \otimes & \times & \times & & & \\ & & & & & & & 0 & 0 & 0 & \otimes \\ & & & & & & & 0 & 0 & \otimes & \times \\ & & & & & & & 0 & \otimes & \times & \times \\ & & & & & & & \otimes & \times & \times & \times \end{bmatrix}$$

and

$$
G_n^{(1)} = \begin{bmatrix}
0 & \otimes & & & & & & & \\
& 0 & \otimes & \times & & & & & \\
& \otimes & \times & \times & & & & & \\
& \otimes & \times & \times & \times & & \otimes & & \\
& & & & 0 & \otimes & \times & & \\
& & & & \otimes & \times & \times & & \\
& & & & \otimes & \times & \times & \times & & \otimes \\
& & & & & & 0 & 0 & \otimes & \times \\
& & & & & & 0 & \otimes & \times & \times \\
& & & & & & \otimes & \times & \times & \times \\
& & & & & & \otimes & \times & \times & \times & \times
\end{bmatrix}.
$$

The entries marked $\otimes$ are different from zero. Also, in each block they are all equal.

1.3 An accurate algorithm to compute zeros of FOPs

We will now discuss an algorithm to compute zeros of FOPs. More precisely, we will show how FOPs can be computed in their product representation. Therefore, the polynomial φ_n immediately leads to the zeros $z_1, \ldots, z_n$. Our numerical experiments indicate that our algorithm gives very accurate results.

Remark 1.3.1. Our aim is to present techniques for computing zeros of analytic functions that give very accurate results. As the reader will notice, if we have a choice between several options for a certain part of an algorithm, then we will always choose the option that, in our experience, gives the most accurate results, even if it is the most expensive (though still within the limits of what is reasonable, of course) in terms of number of floating point operations. The emphasis lies on accuracy.

Theorem 1.2.2 can be interpreted in the following way: the zeros of the regular FOP of degree n can be calculated by solving a generalized eigenvalue problem. The following theorem shows that this zero/eigenvalue property holds for all regular FOPs. This will enable us to evaluate regular FOPs in their product representation, which is numerically very stable. The theorem also provides a solution to the problem of how to switch from ordinary moments to modified moments.

Theorem 1.3.1. *Let $t \geq 1$ be a regular index and let $z_{t,1}, \ldots, z_{t,t}$ be the zeros of the regular FOP φ_t. Then the eigenvalues of the pencil $G_t^{(1)} - \lambda G_t$ are given by $\varphi_1(z_{t,1}), \ldots, \varphi_1(z_{t,t})$. In other words, they are given by $z_{t,1} - \mu, \ldots, z_{t,t} - \mu$ where $\mu = s_1/s_0$.*

Proof. The first part of the proof is similar to the proof of Theorem 1.2.2. Define the Hankel matrix $H_t^<$ as $H_t^< := [s_{1+k+l}]_{k,l=0}^{t-1}$. We will first show that the zeros of φ_t are given by the eigenvalues of the pencil $H_t^< - \lambda H_t$. The zeros of φ_t are given by the eigenvalues of its companion matrix C_t. Let λ^* be an eigenvalue of C_t and x a corresponding eigenvector. As H_t is nonsingular, we may conclude that $C_t x = \lambda^* x \Leftrightarrow H_t C_t x = \lambda^* H_t x$. Using (1.14) one can easily verify that $H_t C_t = H_t^<$.

Let A_t be the unit upper triangular matrix that contains the coefficients of $\varphi_0, \varphi_1, \ldots, \varphi_{t-1}$. Then G_t can be factorized as $G_t = A_t^T H_t A_t$. As $\varphi_1(z) = z - \mu$ where $\mu = s_1/s_0$, the matrix $G_t^{(1)}$ is given by $[\langle \varphi_r, z\varphi_s\rangle]_{r,s=0}^{t-1} - \mu G_t$. The matrix $[\langle \varphi_r, z\varphi_s\rangle]_{r,s=0}^{t-1}$ can be written as $A_t^T H_t^< A_t$ and thus $G_t^{(1)} = A_t^T(H_t^< - \mu H_t)A_t$. Now let λ^* be an eigenvalue of the pencil $H_t^< - \lambda H_t$ and x a corresponding eigenvector. Then

$$H_t^< x = \lambda^* H_t x$$
$$\Leftrightarrow (H_t^< - \mu H_t)x = (\lambda^* - \mu)H_t x$$
$$\Leftrightarrow A_t^T(H_t^< - \mu H_t)A_t y = \varphi_1(\lambda^*)A_t^T H_t A_t y \quad \text{if } y := A_t^{-1}x$$
$$\Leftrightarrow G_t^{(1)}y = \varphi_1(\lambda^*)G_t y.$$

This proves the theorem.   □

Corollary 1.3.1. *The eigenvalues of* $G_n^{(1)} - \lambda G_n$ *are given by* $z_1 - \mu, \ldots, z_n - \mu$ *where* $\mu = s_1/s_0$.

Regular FOPs are characterized by the fact that the determinant of a Hankel matrix is different from zero, while inner polynomials correspond to singular Hankel matrices. To decide whether $\varphi_t(z)$ should be defined as a regular FOP or as an inner polynomial, one could therefore calculate the determinant of H_t and check if it is equal to zero. However, from a numerical point of view such a test "is equal to zero" does not make sense. Because of rounding errors (both in the evaluation of $\langle \cdot, \cdot \rangle$ and the calculation of the determinant) we would encounter only regular FOPs. Strictly speaking one could say that inner polynomials are not needed in numerical calculations. However, the opposite is true! Let us agree to call a regular FOP *well-conditioned* if its corresponding Yule-Walker system (1.14) is well-conditioned, and *ill-conditioned* otherwise. To obtain a numerically stable algorithm, it is crucial to generate only well-conditioned regular FOPs and to replace ill-conditioned regular FOPs by inner polynomials. Stable look-ahead solvers for linear systems of equations that have Hankel structure are based on this principle [20, 28, 46]. In this approach the diagonal blocks in G_n are taken (slightly) larger than strictly necessary to avoid ill-conditioned blocks. A disadvantage is that part of the structure of G_n and $G_n^{(1)}$ gets lost, i.e., there will be some additional fill-in.

Our algorithm for calculating $z_1, \ldots, z_n$ proceeds by computing the polynomials $\varphi_0(z), \varphi_1(z), \ldots, \varphi_n(z)$ in their product representation, starting with $\varphi_0(z) \leftarrow 1$ and $\varphi_1(z) \leftarrow z - \mu$. At each step we ask ourselves whether it is numerically feasible to generate the next polynomial in the sequence as a

regular FOP. As the reader will see, there are several ways to find an answer
to this question.

How do we obtain the value of n? Theorem 1.2.1 and Equations (1.12)
and (1.16) imply the following.

Theorem 1.3.2. *Let* $t \geq n$. *Then* $\varphi_t(z_k) = 0$ *for* $k = 1,\ldots,n$ *and*
$\langle z^p, \varphi_t(z) \rangle = 0$ *for all* $p \geq 0$.

The value of n can be determined as follows. Suppose that the algorithm
has just generated a (well-conditioned) regular FOP $\varphi_r(z)$. To check whether
$n = r$, we scan the sequence

$$\left(|\langle (z - \mu)^\tau \varphi_r(z), \varphi_r(z) \rangle| \right)_{\tau=0}^{N-1-r}. \tag{1.23}$$

If all the elements are "sufficiently small," then we conclude that indeed $n = r$
and we stop.

The form $\langle \cdot, \cdot \rangle$ is evaluated via numerical integration along γ. In other
words, it is approximated by a quadrature sum. We assume that this sum
is calculated in the standard way, by adding the terms one by one, in other
words, by forming a sequence of partial sums. We ask the quadrature algorithm not only for an approximation of the integral, say `result`, but also for
the modulus of the partial sum that has the largest modulus, say `maxpsum`.
Then

$$\log_{10} \frac{\texttt{maxpsum}}{|\texttt{result}|}$$

is an estimate for the loss of precision. This information will turn out to be
extremely useful, for example in the stopping criterion.

These considerations lead to the following algorithm.

ALGORITHM

input $\langle \cdot, \cdot \rangle$, ϵ_{stop}
output n, zeros
comment zeros $= \{z_1, \ldots, z_n\}$. We assume that $\epsilon_{\text{stop}} > 0$.
$N \leftarrow \langle 1, 1 \rangle$

if $N == 0$ **then**
   $n \leftarrow 0$; zeros $\leftarrow \emptyset$; **stop**
else
   $\varphi_0(z) \leftarrow 1$
   $\mu \leftarrow \langle z, 1 \rangle / N$; $\varphi_1(z) \leftarrow z - \mu$
   $r \leftarrow 1$; $t \leftarrow 0$
   while $r + t < N$ **do**
      regular $\leftarrow$ it is numerically feasible to generate $\varphi_{r+t+1}(z)$ as
                  a regular FOP   ...[1]
      if regular **then**

```
          generate φ_{r+t+1}(z) as a regular FOP      ...[2]
          r ← r + t + 1; t ← 0
          allsmall ← true; τ ← 0
          while allsmall and (r + τ < N) do
             [ip, maxsum] ← ⟨(z − μ)^τ φ_r(z), φ_r(z)⟩     ...[3]
             ip ← |ip|
             allsmall ← (ip/maxpsum < ε_stop)    ...[4]
             τ ← τ + 1
          end while
          if allsmall then
             n ← r; zeros ← roots(φ_r); stop
          end if
       else
          generate φ_{r+t+1}(z) as an inner polynomial    ...[5]
          t ← t + 1
       end if
    end while
    n ← N; zeros ← roots(φ_N); stop
 end if
```

Comments:

1. Statement [1] is crucial. But how does one decide that it is numerically feasible to generate the next polynomial in the sequence

$$\varphi_0(z), \varphi_1(z), \ldots, \varphi_n(z)$$

as a regular FOP? Suppose that the algorithm has just generated a regular FOP $\varphi_r(z)$. Then (1.15) implies that if r is a regular index, then $r+1$ is a regular index if and only if $\langle \varphi_r(z), \varphi_r(z) \rangle \neq 0$. This suggests the following criterion: if $|\langle \varphi_r(z), \varphi_r(z) \rangle|/\text{maxsum} < \epsilon_{\text{regular}}$, where $\epsilon_{\text{regular}}$ is some small threshold given by the user, then define $\varphi_{r+1}(z)$ as an inner polynomial, else define it as a regular FOP. However, as we will illustrate in the next section, it is very difficult to choose an appropriate value of $\epsilon_{\text{regular}}$.

We prefer to use the following criterion: act as if the next polynomial in the sequence, say $\varphi_t(z)$, is defined as a regular FOP, i.e., compute its zeros by computing the eigenvalues of the pencil $G_t^{(1)} - \lambda G_t$ and then check if these zeros lie sufficiently close to the interior of γ. If so, then define $\varphi_t(z)$ as a regular FOP, else define it as an inner polynomial. The idea behind this strategy is the following. If the matrix G_t is singular, in which case also the matrix H_t is singular of course, then the pencil $G_t^{(1)} - \lambda G_t$ has either a number of eigenvalues at infinity or a number of eigenvalues that may assume arbitrary values. Indeed, by using the structure of the matrices $G_t^{(1)}$ and G_t one can easily prove the following result, which complements Theorem 1.3.1.

Theorem 1.3.3. *Let $t \geq 1$ be an integer, let r be the largest regular index less than or equal to t, and let r' be the smallest regular index greater than t. (Define $r' := +\infty$ if $t \geq n$.) Then the eigenvalues of the pencil $G_t^{(1)} - \lambda G_t$ are given by the eigenvalues of the pencil $G_r^{(1)} - \lambda G_r$ and $t - r$ eigenvalues that may assume arbitrary values if $t < r' - 1$ or $t - r$ eigenvalues $\lambda = \infty$ if $t = r' - 1$.*

Each of these indeterminate eigenvalues corresponds to two corresponding zeros on the diagonals of the generalized Schur decomposition of $G_t^{(1)}$ and G_t. When actually calculated, these diagonal entries are different from zero because of roundoff errors. The quotient of two such corresponding diagonal entries is a spurious eigenvalue. Our strategy is based on the assumption that, if the matrix H_t, and thus also the matrix G_t, is nearly singular, then the computed eigenvalues of the pencil $G_t^{(1)} - \lambda G_t$ that correspond to the eigenvalues that lie at infinity or that may assume arbitrary values, lie far away from the interior of γ.

The reader may object that our criterion is too strict. Indeed, the zeros of the regular FOPs of degree $< n$ need not lie close to the interior of γ, except if the form $\langle \cdot, \cdot \rangle$ is a true (positive definite) inner product, in which case the zeros of the regular FOPs lie in the convex hull of $\{z_1, \ldots, z_n\}$. (This follows from a general result on orthogonal polynomials. See, e.g., Van Assche [118].) Thus it may very well be the case that some of the computed zeros of a well-conditioned regular FOP lie far away from γ, in which case our algorithm decides to define this polynomial as an inner polynomial. In other words, our algorithm may define more inner polynomials than strictly necessary. However, we have done a lot of numerical tests and have found that our strategy leads to very accurate results. Also, compared to the criterion based on inner products of the type $\langle \varphi_r(z), \varphi_r(z) \rangle$, another advantage is that the user doesn't have to supply a threshold such as $\epsilon_{\text{regular}}$.

2. Statement **[2]** means: define $\varphi_{r+t+1}(z)$ as $\varphi_{r+t+1}(z) \leftarrow \prod_{j=1}^{r+t+1}(z - \alpha_j)$. The zeros α_j are computed as $\alpha_j = \mu + \lambda_j$, $j = 1, \ldots, r + t + 1$, where $\lambda_1, \ldots, \lambda_{r+t+1}$ are the eigenvalues of the pencil $G_{r+t+1}^{(1)} - \lambda G_{r+t+1}$, cf. Theorem 1.3.1.

3. In statement **[3]** we use the inner product $\langle (z - \mu)^\tau \varphi_r(z), \varphi_r(z) \rangle$ and not $\langle z^\tau \varphi_r(z), \varphi_r(z) \rangle$ as it is likely that the former leads to more accurate results than the latter. If $\tau \leq r$, then one may also use $\langle \varphi_r(z) \varphi_r(z), \varphi_r(z) \rangle$. In general, if $\tau = \alpha r + \beta$, where $\alpha, \beta \in \mathbb{N}$ with $\beta < r$, then one may use $\langle [\varphi_r(z)]^{\alpha+1} \varphi_\beta(z), \varphi_r(z) \rangle$.

4. Observe that in statement **[4]** we do not compare ip with ϵ_{stop} but take into account the loss of precision as estimated by the quadrature algorithm. We have found this heuristic to be very reliable.

5. In statement [5] one may define $\varphi_{r+t+1}(z)$ as $\varphi_{r+t+1}(z) \leftarrow (z-\mu)\varphi_{r+t}(z)$ or $\varphi_{r+t+1}(z) \leftarrow \varphi_{t+1}(z)\varphi_r(z)$. In our opinion, both versions are to be preferred to the "classical" $\varphi_{r+t+1}(z) \leftarrow z^{t+1}\varphi_r(z)$.

6. Instead of computing μ, the arithmetic mean of the zeros, as $\mu \leftarrow \langle z, 1 \rangle / N$, one can also use the following formula, which may give a more accurate result: $\mu \leftarrow w + \langle z - w, 1 \rangle / N$, where w is a point inside γ, preferably near the centre of the interior of γ.

7. As we represent our FOPs by using the product representation,

$$\varphi(z) = \prod_{\alpha \in \varphi^{-1}(0)} (z - \alpha),$$

the funtion roots $(\cdot)$ is obviously *not* a function that calculates the zeros of a polynomial from its coefficients in the standard monomial basis.

We have implemented our algorithm in Matlab (for disks) and also in Fortran 90 (for rectangular regions). The latter implementation will be presented in more detail in Section 1.5. Numerical examples will be given in Section 1.4 and also in Section 1.5. They will illustrate the effectiveness of our approach.

1.4 Numerical examples

In the following examples we have considered the case that γ is a circle. The computations have been done via Matlab 5 (with floating point relative accuracy $\approx 2.2204 \, 10^{-16}$).

The following integration algorithm is used to approximate the form $\langle \cdot, \cdot \rangle$. Let γ be the circle with centre c and radius ρ. Then

$$\langle \phi, \psi \rangle = \rho \int_0^1 \phi(c + \rho e^{2\pi i\theta}) \psi(c + \rho e^{2\pi i\theta}) \frac{f'(c + \rho e^{2\pi i\theta})}{f(c + \rho e^{2\pi i\theta})} e^{2\pi i\theta} \, d\theta. \qquad (1.24)$$

Since this is the integral of a periodic function over a complete period, the trapezoidal rule is an appropriate quadrature rule. If $F : [0,1] \to \mathbb{C}$ is the integrand in the right-hand side of (1.24), then the q-point trapezoidal rule approximation to $\langle \phi, \psi \rangle$ is given by

$$\langle \phi, \psi \rangle = \int_0^1 F(\theta) \, d\theta \approx \frac{1}{q} \sum_{k=0}^q {}'' F(k/q) =: T_q.$$

The double prime indicates that the first and the last term of the sum are to be multiplied by 1/2. As F is periodic with period one, we may rewrite T_q as

$$T_q = \frac{1}{q} \sum_{k=0}^{q-1} F(k/q).$$

This shows that T_q indeed depends on q (and not $q+1$) points. As

$$T_{2q} = \frac{1}{2}T_q + T_{q\to 2q}$$

where

$$T_{q\to 2q} := \frac{1}{2q}\sum_{k=0}^{q-1} F\left(\frac{2k+1}{2q}\right),$$

successive doubling of q enables us in each step to reuse the integrand values needed in the previous step. In the following examples we started with $q = 16$ and continued doubling q until $|T_{2q}-T_q|$ was sufficiently small. More precisely, if S_q and $S_{q\to 2q}$ denote the modulus of the partial sum of qT_q respectively $2qT_{q\to 2q}$ that has the largest modulus, then our stopping criterion is given by $|T_{2q} - T_q| \le S_{2q}10^{-14}$, where $S_{2q} := \max\{S_q, S_{q\to 2q}\}/2q$.

Lyness and Delves [96] studied the asymptotic behaviour of the quadrature error. They showed that the modulus of the error made by the q-point trapezoidal rule is asymptotically $\mathcal{O}(A^q)$ where $0 \le A < 1$. More precisely,

$$A := \max\left\{\frac{|z_I|}{\rho}, \frac{\rho}{|z_E|}, \frac{\rho}{\rho_s}\right\}$$

where z_I is the zero of f that lies closest to γ and in the interior of γ, z_E is the zero of f that lies closest to γ and in the exterior of γ, and ρ_s is the distance between c and the nearest singularity of f.

Example 1.4.1. Our first example illustrates the importance of shifting the origin in the complex plane to the arithmetic mean of the zeros. It also compares the two strategies that we have proposed to decide whether it is numerically feasible to generate the next polynomial in the sequence $\varphi_0(z), \varphi_1(z), \ldots, \varphi_n(z)$ as a regular FOP. We will see that it is indeed better to act as if the polynomial is a regular FOP, i.e., to compute its zeros by solving the generalized eigenvalue problem of Theorem 1.3.1, and then to check if these zeros lie sufficiently close to the interior of γ. If so, the polynomial is indeed defined as a regular FOP, else it is defined as an inner polynomial.

Suppose that $n = 3$, $z_1 = \epsilon$, $z_2 = \sqrt{3}+i$, $z_3 = \sqrt{3} - i$, and $\nu_1 = \nu_2 = \nu_3 = 1$. That is, suppose that $f(z) = (z - \epsilon)[(z - \sqrt{3})^2 + 1]$. If $\epsilon = 0$, then the Hankel matrix H_2 is exactly singular, i.e., $\varphi_2(z)$ has to be defined as an inner polynomial. We set $\epsilon = 10^{-2}$. Suppose that $\gamma = \{z \in \mathbb{C} : |z| = 3\}$. In the quadrature algorithm, we have evaluated the logarithmic derivative $f'(z)/f(z)$ of $f(z)$ via the formula

$$\frac{f'(z)}{f(z)} = \sum_{k=1}^{n} \frac{\nu_k}{z - z_k}. \tag{1.25}$$

We have taken $\epsilon_{stop} = 10^{-18}$. Our algorithm proceeds as follows. The total number of zeros N is equal to 3. The polynomial $\varphi_0(z)$ is of course defined as a regular FOP, $\varphi_0(z) \leftarrow 1$. The arithmetic mean μ is approximated by

$$1.158033871706381\,e{+}00 \; + \; i \;\; 1.526556658859590\,e{-}16.$$

The polynomial $\varphi_1(z)$ is also defined as a regular FOP, $\varphi_1(z) \leftarrow z - \mu$. The inner product $\langle \varphi_1(z), \varphi_1(z) \rangle$ is equal to 0.02303. To take into account the loss of precision, we divide by S_{2q} to obtain its scaled counterpart, which is equal to 0.01231. Is this that small that we should define $\varphi_2(z)$ as an inner polynomial? It seems not, and we decide to define $\varphi_2(z)$ as a regular FOP. Its zeros are approximated by

$$1.158072473165547\,e{+}00 \; + \; i \;\; 1.513258564730580\,e{-}16$$
$$2.000049565733722\,e{+}02 \; + \; i \;\; 3.487413819470906\,e{-}12$$

Note how large the second zero is! The inner product $\langle \varphi_2(z), \varphi_2(z) \rangle$ is equal to 910.504. Its scaled counterpart is 0.01214. We decide to define $\varphi_3(z)$ as a regular FOP. Its zeros are approximated by

$$1.732050807571817\,e{+}00 \; - \; i \;\; 1.000000000004209\,e{+}00$$
$$1.000000000661760\,e{-}02 \; + \; i \;\; 3.913802997325630\,e{-}12$$
$$1.732050807566777\,e{+}00 \; + \; i \;\; 1.000000000002807\,e{+}00$$

As $N = 3$, we may stop. The relative errors of the approximations for the zeros of f are $\mathcal{O}(10^{-12})$, except for the zero that approximates $z_1 = \epsilon$, which has a relative error of $\mathcal{O}(10^{-10})$. The absolute errors are $\mathcal{O}(10^{-12})$. The relative errors of the approximations for the multiplicities are $\mathcal{O}(10^{-11})$.

As one of the zeros of $\varphi_2(z)$ lies far away from the interior of γ, we should decide to define $\varphi_2(z)$ as an inner polynomial. Surprisingly, this does not improve the accuracy of the results. However, let us see what happens if we first shift the origin to μ, or, equivalently, if we consider the circle $\gamma = \{ z \in \mathbb{C} \, : \, |z - \mu| = 2 \}$. Note that we change both the centre and the radius of γ. By defining $\varphi_2(z)$ as a regular FOP, the accuracy of the results does not improve. However, if we define $\varphi_2(z)$ as an inner polynomial, the relative errors of the approximations for the zeros of f are $\mathcal{O}(10^{-16})$, except for the zero that approximates $z_1 = \epsilon$, which has a relative error of $\mathcal{O}(10^{-14})$. The absolute errors are $\mathcal{O}(10^{-16})$. The relative errors of the approximations for the multiplicities are $\mathcal{O}(10^{-16})$. In other words, the results are indeed much better. $\diamond$

Example 1.4.2. Let $f(z) = e^{3z} + 2z \cos z - 1$ and $\gamma = \{ z \in \mathbb{C} \, : \, |z| = 2 \}$. We set $\epsilon_{\text{stop}} = 10^{-18}$. Our algorithm finds that $N = 4$. It defines $\varphi_0(z)$ and $\varphi_1(z)$ as regular FOPs. From the eigenvalues of the pencil $G_2^{(1)} - \lambda G_2$ it concludes that $\varphi_2(z)$ would have a zero of modulus ≈ 43 in case $\varphi_2(z)$ is defined as a regular FOP. Thus the algorithm decides to define $\varphi_2(z)$ as an inner polynomial. The polynomials $\varphi_3(z)$ and $\varphi_4(z)$ are defined as regular FOPs. Our algorithm concludes that $n = 4$. The computed approximations for the zeros of f are given by

$$-1.844233953262213\,e{+}00 \; - \; i \;\; 1.106288924192872\,e{-}16$$
$$5.308949302929297\,e{-}01 \; + \; i \;\; 1.331791876751121\,e{+}00$$

```
 5.308949302929303 e-01 - i 1.331791876751121 e+00
-5.412337245047638 e-15 + i 3.762630283199076 e-16
```

The corresponding approximations for the multiplicities are

```
1.000000000000001 e+00 + i 9.279422312879846 e-17
1.000000000000001 e+00 - i 2.415808667423342 e-15
1.000000000000001 e+00 + i 1.187431378902999 e-15
9.999999999999974 e-01 + i 1.142195850770565 e-15
```

By refining the approximations for the zeros of f via Newton's method, we find that they have a relative error of $\mathcal{O}(10^{-16})$, except for the approximation of $z_4 = 0$, which has an absolute error of $\mathcal{O}(10^{-15})$. If $\varphi_2(z)$ is defined as a regular FOP, the errors are $\mathcal{O}(10^{-13})$.   ◊

Example 1.4.3. Suppose that $f(z) = z^2(z-1)(z-2)(z-3)(z-4) + z \sin z$ and $\gamma = \{z \in \mathbb{C} : |z| = 5\}$. Note that f has a double zero at the origin. Our algorithm finds that $N = 6$. It defines $\varphi_0(z)$, $\varphi_1(z)$, $\varphi_2(z)$, $\varphi_3(z)$, $\varphi_4(z)$ and $\varphi_5(z)$ as regular FOPs. For $k = 2, 3, 4$ and 5, the scaled counterparts of $|\langle \varphi_k(z), \varphi_k(z) \rangle|$ are given by

```
7.040331724952680 e-02
1.910625118197985 e-03
1.236575744513765 e-05
2.425617684377941 e-15
```

If we set ϵ_{stop} to a value that is larger than $2.5\,10^{-15}$, then the algorithm stops. It decides (correctly) that $n = 5$. The computed approximations for the zeros of f are given by

```
2.853939307101427 e-12 + i 1.218663779565894 e-12
1.189065889993786 e+00 + i 1.174492347177040 e-10
1.728434986506658 e+00 + i 1.587636971134256 e-10
3.019907328131211 e+00 + i 1.757549887398162 e-11
4.030381916062330 e+00 + i 7.769722658005202 e-13
```

The corresponding approximations for the multiplicities are

```
2.000000000021947 e+00 + i 9.312570667099433 e-12
1.000000000957614 e+00 + i 4.352560366652006 e-10
9.999999990867526 e-01 - i 4.117654248790392 e-10
9.999999999418555 e-01 - i 2.906361792434579 e-11
9.999999999918315 e-01 - i 3.739564528915041 e-12
```

By refining the approximations for the zeros iteratively via Newton's method, we find that the absolute errors are $\mathcal{O}(10^{-12})$, $\mathcal{O}(10^{-12})$, $\mathcal{O}(10^{-12})$, $\mathcal{O}(10^{-14})$ and $\mathcal{O}(10^{-16})$, respectively. The refined approximations for the zeros are

```
2.853939307034829 e-12 + i 1.218663779496334 e-12
1.189065890003747 e+00 + i 1.218863825423787 e-10
```

$$
\begin{array}{l}
\texttt{1.728434986503592e+00 + i 1.573384095073740e-10} \\
\texttt{3.019907328131241e+00 + i 1.759027974661844e-11} \\
\texttt{4.030381916062330e+00 + i 7.768170803990681e-13}
\end{array}
$$

If we set ϵ_{stop} to a value that is smaller than $2.4\,10^{-15}$, then the algorithm continues. It defines $\varphi_6(z)$ as a regular FOP and stops. The computed approximations for the zeros of f are now given by

$$
\begin{array}{l}
\texttt{-3.412647942013791e-11 - i 3.671305864695379e-12} \\
\texttt{ 1.189065887181205e+00 - i 7.212695548387926e-11} \\
\texttt{ 1.728434983085149e+00 - i 2.216255731479441e-11} \\
\texttt{ 3.019907327801388e+00 + i 1.210604626933557e-11} \\
\texttt{ 4.030381916045125e+00 + i 1.488578696458235e-12} \\
\texttt{-2.858307137641358e+00 + i 1.433406287105723e+00}
\end{array}
$$

whereas the corresponding approximations for the multiplicities are

$$
\begin{array}{l}
\texttt{1.999999999750989e+00 - i 1.982896140226239e-11} \\
\texttt{9.999999911197371e-01 - i 1.399505295104063e-10} \\
\texttt{1.000000008572735e+00 + i 1.723707922538161e-10} \\
\texttt{1.000000000489271e+00 - i 7.095346599149408e-12} \\
\texttt{1.000000000067249e+00 - i 5.486513691338396e-12} \\
\texttt{2.023577344691327e-14 - i 9.441050659615426e-15}
\end{array}
$$

Observe that we find more or less the same approximations for the zeros as before and also a spurious "zero," cf. Theorem 1.3.3. Fortunately, the presence of spurious zeros can be easily detected, as their corresponding "multiplicities" are equal to zero. This can be explained as follows. The Vandermonde matrix that corresponds to the calculated approximations for the zeros will almost surely (i.e., with probability one) be nonsingular and therefore the system for the multiplicities will have only one solution, which gives the true zeros of f their correct corresponding multiplicity and the spurious ones multiplicity zero.   ◊

Example 1.4.4. Let $f(z) = z^2(z-2)^2[e^{2z}\cos z + z^3 - 1 - \sin z]$ and $\gamma = \{z \in \mathbb{C} : |z| = 3\}$. Note that f has a triple zero at the origin and a double zero at $z = 2$. Our algorithm finds that $N = 8$. It defines $\varphi_0(z)$, $\varphi_1(z)$, $\varphi_2(z)$, $\varphi_3(z)$, $\varphi_4(z)$ and $\varphi_5(z)$ as regular FOPs. For $k = 2, 3, 4$ and 5, the scaled counterparts of $|\langle \varphi_k(z), \varphi_k(z) \rangle|$ are given by

$$
\begin{array}{l}
\texttt{1.515262455417321e-01} \\
\texttt{3.291787369922850e-02} \\
\texttt{2.663301879352920e-03} \\
\texttt{5.361771800010522e-15}
\end{array}
$$

If we set ϵ_{stop} to a value that is larger than $5.4\,10^{-15}$, then the algorithm stops. It decides (correctly) that $n = 5$. The computed approximations for the zeros of f are given by

```
-4.607141197276816 e-01 + i 6.254277693477422 e-01
-4.607141197280875 e-01 - i 6.254277693470294 e-01
 2.653322006551662 e-12 + i 8.232702952841388 e-13
 2.000000000000728 e+00 + i 2.133638388712091 e-13
 1.664682869749229 e+00 + i 9.605975127915706 e-13
```

The corresponding approximations for the multiplicities are

```
1.000000000003826 e+00 + i 5.208513390226470 e-12
1.000000000005729 e+00 - i 1.653703539980123 e-12
2.999999999991895 e+00 - i 3.202939915103751 e-12
1.999999999987495 e+00 - i 3.449339571824102 e-12
1.000000000011054 e+00 + i 3.097374215634804 e-12
```

By refining the approximations for the zeros iteratively via Newton's method, we find that the absolute errors are $\mathcal{O}(10^{-14})$, $\mathcal{O}(10^{-12})$, $\mathcal{O}(10^{-12})$, $\mathcal{O}(10^{-13})$ and $\mathcal{O}(10^{-12})$, respectively.

If we set ϵ_{stop} to a value that is smaller than 10^{-15}, then the algorithm continues until the very end. For $k = 6$ and 7, the scaled counterparts of $|\langle \varphi_k(z), \varphi_k(z) \rangle|$ are given by

```
3.073186423911271 e-15
1.658776045256565 e-15
```

The algorithm defines $\varphi_6(z)$ and $\varphi_7(z)$ as regular FOPs. Then it defines $\varphi_8(z)$ as an inner polynomial and stops. The computed approximations for the zeros of f are now given by

```
-4.607141197287676 e-01 + i 6.254277693479825 e-01
-4.607141197287121 e-01 - i 6.254277693478455 e-01
 6.096234628216735 e-13 + i 4.562446979887560 e-14
 2.000000000000117 e+00 - i 2.873008259688547 e-13
 1.664682869746416 e+00 - i 1.002467422184267 e-12
-2.764701251474829 e+00 + i 1.339901195840935 e+00
 1.042158257408570 e+00 - i 3.110403055104297 e+00
 5.929068287859467 e-01 - i 9.822871682718670 e-17
```

whereas the corresponding approximations for the multiplicities are

```
1.000000000000205 e+00 + i 1.226534018772486 e-12
1.000000000000592 e+00 - i 1.131203882923728 e-12
2.999999999999770 e+00 - i 2.952132467539878 e-13
1.999999999997560 e+00 + i 4.325897066507562 e-12
1.000000000001872 e+00 - i 4.180464412170987 e-12
5.712723743305506 e-16 + i 2.376958372514238 e-16
4.180089925492345 e-16 - i 2.883533010129535 e-16
1.914114074394907 e-13 - i 1.417681479033546 e-13
```

As discussed in Example 1.4.3, the approximations for the multiplicities enable us to locate spurious zeros.   ◊

Example 1.4.5. Let us consider the Wilkinson polynomial of degree 10, $f(z) = \prod_{k=1}^{10}(z - k)$. Suppose that $\gamma = \{z \in \mathbb{C} : |z| = 11\}$. We have evaluated the logarithmic derivative of f via the formula (1.25). By using Theorem 1.2.2 we obtain the following approximations for the zeros of f:

```
1.000069039770840 e+00 - i 6.902850482141518 e-05
2.004976265975782 e+00 - i 3.428732045109014 e-03
3.060983206011711 e+00 - i 2.595941486831997 e-02
4.251792296513047 e+00 - i 6.062429488533006 e-02
5.546923068519217 e+00 - i 6.702269982182261 e-02
6.815527607409872 e+00 - i 4.182297169281859 e-02
7.960662733297854 e+00 - i 1.296711240552063 e-02
8.997081596675446 e+00 - i 1.314415641505938 e-03
9.999960580756190 e+00 - i 2.207636166991653 e-05
3.755000277916281 e+00 - i 4.101268365758313 e+00
```

The absolute errors of the approximations for the multiplicities are $\mathcal{O}(10^{-1})$.

Now let us see how our algorithm performs. We set $\epsilon_{\text{stop}} = 10^{-18}$. It finds that $N = 10$. It defines $\varphi_0(z), \varphi_1(z), \ldots, \varphi_9(z)$ as regular FOPs. For $k = 2, 3, \ldots, 9$, the scaled counterparts of $|\langle \varphi_k(z), \varphi_k(z) \rangle|$ are given by

```
3.816240840811839 e-02
1.299688508906636 e-03
3.553824638723713 e-05
8.039698366555619 e-07
1.469066688160238 e-08
2.010063990239241 e-10
1.220433804961808 e-12
9.741759723349369 e-16
```

This clearly indicates that the problem is very ill-conditioned. The algorithm defines $\varphi_{10}(z)$ as an inner polynomial and stops. The relative errors of the approximations for the zeros are given by

```
1.005577196493408 e-04
2.863610448143864 e-03
2.009371359668208 e-02
5.876683150395577 e-02
9.999999999999894 e-02
8.141857700057402 e-02
3.100882100301538 e-02
6.609036435112427 e-03
5.202783593342222 e-04
7.598125455914798 e-06
```

respectively. However, if we consider the circle $\gamma = \{z \in \mathbb{C} : |z - 5.5| = 5.5\}$, then the following happens. The algorithm defines only regular FOPs. For $k = 2, 3, \ldots, 9$, the scaled counterparts of $|\langle \varphi_k(z), \varphi_k(z) \rangle|$ are now given by

```
1.120889480330385 e+00
3.130941249560411 e-01
6.391288244829463 e-02
9.005452570225671 e-03
1.142227363432899 e-03
1.122128881438772 e-04
2.425547309622016 e-06
2.149981073624070 e-08
```

The relative errors of the approximations for the zeros are given by

```
8.362431354624777 e-12
4.684075580626331 e-10
6.568404577352624 e-09
3.252872101214176 e-08
6.598954186243715 e-08
5.800798368415380 e-08
2.196536348456049 e-08
3.331673425077182 e-09
1.672314153254196 e-10
1.651045356918564 e-12
```

whereas the absolute errors of the approximations for the multiplicities are

```
4.518653295430768 e-11
3.537203913804699 e-09
4.763707040887607 e-08
1.808818111196228 e-07
1.605226012463829 e-07
1.316877594500569 e-07
1.948007501440231 e-07
6.042232266569582 e-08
5.351870471708607 e-09
9.475094893599826 e-11
```

Example 1.4.6. Recently, Engelborghs, Luzyanina and Roose [41] have used our algorithm to compute all the zeros of

$$f(z) = a + bz + z^2 - hz^2 e^{-\tau z},$$

where $a = 1$, $b = 0.5$, $h = -0.82465048736655$ and $\tau = 6.74469732735569$, that lie in the rectangular region

$$\{z \in \mathbb{C} : -0.3 \leq \operatorname{Re} z \leq 0.1, \quad -24.7 \leq \operatorname{Im} z \leq 24.7\}.$$

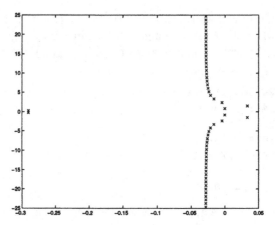

Fig. 1.3. Zeros of $a + bz + z^2 - hz^2e^{-\tau z}$ where $a = 1$, $b = 0.5$, $h = -0.82465048736655$ and $\tau = 6.74469732735569$.

The results are shown in Figure 1.3. These zeros determine the stability of a steady state solution of a neutral functional differential equation. For the given values of the parameters a, b, h and τ, this steady state solution is also a Hopf bifurcation point. The authors have transformed the computed zeros into the Floquet multipliers of the corresponding emanating periodic solution. This has enabled them to compare these "exact" Floquet multipliers with the approximations obtained via their approach.   ◇

Note 1.4.1. As our algorithm not only obtains approximations for the zeros but also the corresponding multiplicities, we can use the modified Newton's method

$$z_k^{(\alpha+1)} = z_k^{(\alpha)} - \nu_k \frac{f(z_k^{(\alpha)})}{f'(z_k^{(\alpha)})}, \qquad \alpha = 0, 1, 2, \ldots,$$

to refine the approximations for the zeros.

We will give more numerical examples in Section 1.5 and in Chapter 2.

1.5 The software package ZEAL

We have implemented the algorithm that we have presented in Section 1.3 for the case that the curve γ is a rectangle whose edges are parallel to the coordinate axes. Our package is called ZEAL ('ZEros of AnaLytic functions') and is written in Fortran 90 [91]. Numerical approximations for the integrals along γ are computed via the quadrature package QUADPACK [109].

Once the user has specified the rectangle γ, the analytic function f, its first derivative f' and the value of the parameter M (i.e., the number of zeros

that are to be calculated simultaneously, cf. our discussion of the method of
Delves and Lyness in Section 1.1), he/she can ask ZEAL to compute only
the total number of zeros of f that lie inside γ, to isolate subregions of the
interior of γ that contain at most M zeros, to compute all the zeros of f that
lie inside γ or to compute only a specified number of zeros (together with
their respective multiplicities). The results can be written to separate files,
etc. All these options will be discussed in detail below.

1.5.1 The approach taken by ZEAL

Given a rectangle whose edges are parallel to the coordinate axes and a
positive integer M, we take the following approach:

- We calculate the total number of zeros that lie inside this rectangle.
- Via consecutive subdivisions we obtain a set of subrectangles, each of which
  contains at most M zeros (counting multiplicities).
- For each of these subrectangles, we calculate approximations for the zeros
  that lie inside it, together with their respective multiplicities.
- The approximations for the zeros are refined iteratively via the modified
  Newton's method.

As the function f may have zeros on the boundary of the rectangular box
specified by the user, ZEAL starts by perturbing this box. For this purpose a
tolerance is used that is taken to be proportional to a power of the machine
precision, for example 10 times the square root of the machine precision. The
box is then slightly enlarged in an asymmetrical way. The reason for this
asymmetric perturbation is to lower the possibility of having a zero close to
or on any boundary of the consecutive subdivisions (see also the footnote
below). For example, if the starting box is symmetric with respect to the
imaginary axis, then the inner boundary at the first subdivision will pass
through any imaginary zeros of f.

The total number of zeros of f is computed as follows. Suppose that the
rectangle γ has left lower vertex (x_0, y_0) and that its edges have length h_1
and h_2. In other words, suppose that γ is the boundary of

$$[x_0, x_0 + h_1] \times [y_0, y_0 + h_2].$$

Let $W^\star := \{ (x, y) \in \mathbb{R}^2 : x + iy \in W \}$ and $u, v : W^\star \to \mathbb{R}$, with

$$u(x, y) := \operatorname{Re} f(x + iy) \qquad \text{and} \qquad v(x, y) := \operatorname{Im} f(x + iy).$$

Define the functions ψ and φ in $W^\star$ as

$$\psi(x, y) := \frac{u(x, y)\, v_x(x, y) - v(x, y)\, u_x(x, y)}{[u(x, y)]^2 + [v(x, y)]^2}$$

and

$$\varphi(x,y) := \frac{u(x,y)\,v_y(x,y) - v(x,y)\,u_y(x,y)}{[u(x,y)]^2 + [v(x,y)]^2}.$$

By considering the real part of the integral in (1.1), one can easily verify that

$$N = h_1\,(I_1 - I_3) + h_2\,(I_2 - I_4)$$

where

$$I_1 := \frac{1}{2\pi}\int_0^1 \psi(x_0 + th_1, y_0)\,dt$$

$$I_2 := \frac{1}{2\pi}\int_0^1 \varphi(x_0 + h_1, y_0 + th_2)\,dt$$

$$I_3 := \frac{1}{2\pi}\int_0^1 \psi(x_0 + th_1, y_0 + h_2)\,dt$$

$$I_4 := \frac{1}{2\pi}\int_0^1 \varphi(x_0, y_0 + th_2)\,dt.$$

In other words, the integral is written as a sum of four integrals, where each integral corresponds to one of the edges of the rectangular region. Approximations for these integrals are calculated via the adaptive integrator DQAG from QUADPACK. A zero near one of the edges of the rectangle causes the integrand of the corresponding integral to have a "peak." The closer the zero lies to the edge, the sharper this peak is. If the zero lies on the edge, then the integral is divergent. DQAG uses adaptive strategies that enable it to cope with such peaks efficiently. However, if a zero lies too close to an edge (the corresponding peak is too sharp), then DQAG warns us that it has problems in calculating the integral. Our algorithm then slightly moves this edge and restarts. By enlarging the user's box, we may of course include additional zeros. We have decided not to discard any of these zeros ourselves. Rather, we provide the user with the box that eventually has been considered, all the zeros that lie inside this box, and leave it to him/her to filter out unwanted zeros.

If the starting box (as perturbed by ZEAL) contains less than M zeros (counting multiplicities), then approximations for these zeros and the values of their respective multiplicities are computed via our implementation of the algorithm that we have presented in Section 1.3. Otherwise, the longest edges of the box are halved and the box is subdivided into two equal boxes. The number of zeros in each of these boxes is calculated via numerical integration. If DQAG detects a zero near the inner edge, then this edge is shifted, a process that results in an asymmetric subdivision of the box.[1] Then the

---

[1] QUADPACK uses certain heuristic strategies. They work very well but nevertheless can fail. For example, we have observed that in case the function has several

two smaller boxes are examined. A box that does not contain any zero is abandoned. If a box contains less than M zeros, then approximations for these zeros are calculated, together with their respective multiplicities. A box that contains more than M zeros is subdivided again. This process is repeated until a set of boxes has been found, each of which contains at most M zeros, and approximations for all these zeros as well as the values of their respective multiplicities have been computed. The approximations for the zeros are then refined via the modified Newton's method, which takes into account the multiplicity of a zero and converges quadratically.

1.5.2 The structure of ZEAL

The package ZEAL contains about 6500 lines of code including comments. It is written in Fortran 90 and has been thoroughly tested on various UNIX machines.

ZEAL consists of 11 parts: the main program Main and the modules

- Precision_Module, in which the user can specify the precision to which the floating point calculations are to be done,
- Zeal_Module, which contains the main subroutine ZEAL and also the subroutines CHECK_INPUT and ERROR_EXIT,
- Zeros_Module, which contains the subroutines APPROXIMATE and INPROD,
- Refine_Module, which contains the subroutines REFINE and NEWTON,
- Split_Module, which contains the subroutines INBOX and SPLITBOX,
- and finally the modules Error_Module, Quad_Module, Zeal_Input_Module, Function_Input_Module and Integration_Input_Module.

ZEAL also requires the subroutine DQAG from the package QUADPACK [109] and a number of subprograms from the BLAS and LAPACK libraries [8].

The user can specify the values of the input parameters by editing the three modules

        Zeal_Input_Module
        Function_Input_Module
        Integration_Input_Module

This will be discussed in Subsection 1.5.3.

The main program Main has the following form:

---

zeros very close (say, at a distance less than 10^{-7}) to the boundary, then DQAG may compute the wrong total number of zeros without giving any warning message. Therefore, we advise the user to make sure (as much as possible) that the zeros are not symmetrical with respect to the boundary. In particular, as many functions have zeros on the coordinate axes, we suggest to consider boxes that are asymmetrical with respect to these axes.

```
PROGRAM Main

USE Precision_Module
USE Zeal_Module

IMPLICIT NONE

INTEGER :: TOTALNUMBER, DISTINCTNUMBER, REFINEDNUMBER
INTEGER, DIMENSION(:), POINTER          :: MULTIPLICITIES
LOGICAL, DIMENSION(:), POINTER          :: REFINEMENT_OK
COMPLEX(KIND=DP), DIMENSION(:), POINTER :: ZEROS, FZEROS

CALL ZEAL(TOTALNUMBER,DISTINCTNUMBER,ZEROS,FZEROS,  &
          MULTIPLICITIES,REFINEDNUMBER,REFINEMENT_OK)

END PROGRAM Main
```

The subroutine ZEAL returns the total number of zeros of the given function that lie inside the given rectangular region, the number of mutually distinct zeros, the refined approximations for the zeros and the values that the function takes at these points, the corresponding multiplicities, the number of approximations for the zeros (as computed by the subroutine APPROXIMATE) that ZEAL has been able to refine successfully via the modified Newton's method, and finally for each computed zero a logical variable that indicates whether this refinement procedure has been successful or not.

In the design of ZEAL we have followed the recommendations for precision level maintenance described by Buckley [23]. The parameter DP that appears in the declaration of the variables ZEROS and FZEROS is defined in Input,

```
INTEGER, PARAMETER :: DP = SELECTED_REAL_KIND(15,70)
```

It determines the precision to which all the floating point calculations are to be done. On the computers that we have used, its current value corresponds to Fortran 77's DOUBLE PRECISION.[2]

Let us briefly describe the various parts of ZEAL.

The subroutine INBOX calculates the total number of zeros that lie inside the rectangular box given by the user. If some of the zeros lie too close to the

---

[2] This observation is important for the following reason. As documented in its makefile, ZEAL uses certain Fortran 77 routines from QUADPACK, BLAS and LAPACK. To enable the user to compile the necessary routines from BLAS and LAPACK in case these libraries are not available on his/her computer system, we have included them with our distribution of ZEAL. However, we have included only the DOUBLE PRECISION version of these routines and hence they should be replaced by the corresponding SINGLE PRECISION routines in case a change to DP requires this. The same holds for the subroutine DQAG from QUADPACK.

boundary of this box and the quadrature routine DQAG fails, then INBOX perturbs the box slightly and enlarges it.

The subroutine SPLITBOX takes a box and splits it into two boxes. A symmetric splitting, which proceeds by halving the longest edges, is tried first. If the calculation of the integral along the inner edge fails, then it is assumed that some of the zeros lie too close to this edge and the inner edge is shifted.

The subroutine APPROXIMATE contains our implementation of the algorithm that we have presented in Section 1.3. The symmetric bilinear form (1.12) is evaluated via the subroutine INPROD.

The subroutine NEWTON contains our implementation of the modified Newton's method. The subroutine REFINE calls NEWTON to refine the approximations for the zeros that APPROXIMATE has computed. If NEWTON fails, then REFINE tries again from a nearby point. If after eight attempts NEWTON still fails, then REFINE indicates that it has been unable to refine the given approximation successfully.

The subroutine ZEAL forms the main part of the package. ZEAL starts by calling CHECK_INPUT to check if the input parameters specified by the user are proper. Next it calls INBOX. If there are no zeros inside the user's box, then the program stops. If there are less than M zeros inside the box (where the value of M can be specified in Zeal_Input_Module), then APPROXIMATE and REFINE are called. Else, the box is given to SPLITBOX. The two boxes returned by SPLITBOX are examined. A box that does not contain any zero is abandoned. A box that contains less than M zeros is given to APPROXIMATE and REFINE. A box that contains more than M zeros is put in a list. Then ZEAL takes a next box from this list and calls SPLITBOX. This procedure is repeated until all the zeros have been computed, together with their respective multiplicities.

The program execution terminates normally after the completion of its task. This type of termination is indicated by the value 1 of the variable INFO, which is a global variable declared in the module Error. If the value of this parameter is different from 1, then the termination of the program is abnormal. The cases of abnormal termination are the following:

INFO=0 Improper input parameters.

INFO=2 The procedure for the calculation of the total number of zeros has failed.

INFO=3 The procedure for the isolation of the zeros has failed.

INFO=4 The procedure for the computation of the zeros has failed.

This concludes our discussion of the structure of ZEAL.

1.5.3 ZEAL's user interface

The user can specify the input parameters by editing three different files. (This splitting was done to speed up the recompilation in case only a few parameters are changed.)

In the module `Zeal_Input_Module` the following parameters have to be set:

LV a real array of length 2 that contains the x- and y-coordinates of the left lower vertex of the rectangle that is to be examined.

H a real array of length 2 that specifies the size of this rectangle along the x- and y-direction.

M an integer that determines the maximum number of zeros (counting multiplicities) that are considered within a subrectangle. M has to be larger than the maximum of the multiplicities of the zeros. A recommended value is 5.

ICON an integer in $\{1, \ldots, 4\}$ that specifies which calculations are to be done:
   1. calculation of the total number of zeros, only,
   2. calculation of the total number of zeros and isolation of a set of subrectangles, each of which contains at most M zeros,
   3. calculation of the total number of zeros and computation of all the zeros, together with their respective multiplicities,
   4. calculation of the total number of zeros and computation of NR zeros, together with their respective multiplicities.

Note that if ICON=4, the user must also supply the desired number of zeros NR. In the other cases (ICON=1,2,3) a value of NR may be supplied but it will not be used by the package.

VERBOSE a logical variable. ZEAL is allowed to print information (concerning the user's input and the computed results) if and only if VERBOSE is equal to .TRUE.

FILES a logical variable. If FILES is set equal to .TRUE. then ZEAL generates the files `zeros.dat` and `mult.dat`. They contain the computed approximations for the zeros as well as their respective multiplicities. ZEAL also writes the file `fzeros.dat`, which contains the values that the function takes at the computed approximations for the zeros.

IFAIL an integer that determines how errors are to be handled. We follow the NAG convention:
   1. *soft silent error*—control is returned to the calling program.
  -1. *soft noisy error*—an error message is printed and control is returned to the calling program.
   0. *hard noisy error*—an error message is printed and the program is stopped.

These parameters determine the geometry of the rectangular region that is to be considered and the type of calculation that ZEAL will perform.

In the module `Integration_Input_Module` the following parameters have to be set:

NUMABS a real variable that determines the absolute accuracy to which the integrals that calculate the number of zeros are to be evaluated. In case NUMABS = 0.0_DP, only a relative criterion is used.

NUMREL a real variable that determines the relative accuracy to which the integrals that calculate the number of zeros are to be evaluated. In case NUMREL = 0.0_DP, only an absolute criterion is used.

If NUMABS and NUMREL are both too small, then the numerical integration may be time-consuming. If they are both too large, then the calculated number of zeros may be wrong. The default values of NUMABS and NUMREL are 0.07_DP and 0.0_DP, respectively. These variables are used by QUADPACK.

INTABS a real variable that determines the absolute accuracy to which the integrals that are used to compute approximations for the zeros are to be calculated. If INTABS = 0.0_DP, then only a relative criterion will be used.

INTREL a real variable that determines the relative accuracy to which the integrals that are used to compute approximations for the zeros are to be calculated. If INTREL = 0.0_DP, then only an absolute criterion will be used.

If INTABS and INTREL are both too small, then the numerical integration may be time-consuming. If they are both too large, then the approximations for the zeros may be very inaccurate and Newton's method, which is used to refine these approximations (see NEWTONZ and NEWTONF), may fail. The default values of INTABS and INTREL are 0.0_DP and 1.0E-12_DP, respectively. These variables are used by QUADPACK.

EPS_STOP a real variable that is used in the stopping criterion that determines the value of n, the number of mutually distinct zeros. If EPS_STOP is too large, then the computed value of n may be smaller than the actual number of distinct zeros. If EPS_STOP is too small, then the computed value of n may be larger than the actual number of distinct zeros, especially in case the function has many multiple zeros. A recommended value is 1.0E-08_DP.

These parameters are related to numerical integration.

Finally, in the module `Function_Input_Module` the user has to specify two parameters that are used in the stopping criteria for Newton's method, the function f whose zeros ZEAL has to compute as well as its first derivative.

He or she also has to give some information about the analyticity of f inside the considered region.

The parameters used to control the Newton's process are the following:

NEWTONZ and NEWTONF these real variables should be specified in case ICON = 3 or 4. They are used as follows. The modified Newton's method, which takes into account the multiplicity of a zero and converges quadratically, is used to refine the calculated approximations for the zeros. The iteration stops if the relative distance between two successive approximations is at most NEWTONZ or the absolute value of the function at the last approximation is at most NEWTONF or if a maximum number of iterations (say, 20) is exceeded.

The considered function f and its first derivative f' should be defined via the subroutine FDF, which takes the following form:

```
SUBROUTINE FDF(Z,F,DF)

COMPLEX(KIND=DP), INTENT(IN)   :: Z
COMPLEX(KIND=DP), INTENT(OUT)  :: F, DF

F = ...
DF = ...

END SUBROUTINE FDF
```

If any cases where f is not analytic are known, they have to be specified using the logical function VALREG. Given a rectangular region specified by its left lower vertex and the sizes of its edges, VALREG decides whether the function f is analytic inside this region or not. ZEAL uses this information to decide whether it is allowed to move the edge of a box or not. VALREG has the following form:

```
FUNCTION VALREG(LV,H)

LOGICAL VALREG
REAL(KIND=DP), INTENT(IN) :: LV(2), H(2)

VALREG = ...

END FUNCTION VALREG
```

For example, if f is analytic in the entire complex plane, then one may use the statement

```
VALREG = .TRUE.
```

If f has a branch cut along the non-positive real axis, then one may write

```
VALREG = .NOT. ( LV(2)*(LV(2)+H(2)) <= 0.0_DP .AND.  &
               LV(1) <= 0.0_DP )
```

This concludes our discussion of ZEAL's user interface.

1.5.4 A few examples of how to use ZEAL

We will now discuss a few numerical examples.

Example 1.5.1. Suppose that $f(z) = e^{3z} + 2z\cos z - 1$ and that

$$W = \{ z \in \mathbb{C} : -2 \leq \operatorname{Re} z \leq 2, \quad -2 \leq \operatorname{Im} z \leq 3 \}.$$

In other words, W is the rectangular region $[-2, 2] \times [-2, 3]$. Therefore, we have to define the input parameters LV and H as

LV = (/-2.0_DP,-2.0_DP/) and H = (/ 4.0_DP, 5.0_DP/).

We set M = 5. The logical variables VERBOSE and FILES are set to .TRUE. We start by calculating only the total number of zeros, ICON = 1. ZEAL outputs the following.

```
This is ZEAL.

Input:

LV    =    -2.00000000000000      -2.00000000000000
H     =     4.00000000000000       5.00000000000000

M     =    5
ICON  =    1

FILES =    T

Results:

The following box has been considered:

LV =  -2.00000016391277      -2.00000019371510
H  =   4.00000035762787       5.00000041723251

Total number of zeros inside this box        =      4
```

The function has four zeros inside the given box. We now ask ZEAL to compute approximations for all these zeros, ICON=3.

This is ZEAL.

Input:

```
LV   =    -2.00000000000000        -2.00000000000000
H    =     4.00000000000000         5.00000000000000


M    =   5
ICON =   3

FILES =   T
```

Results:

The following box has been considered:

```
LV = -2.00000016391277        -2.00000019371510
H  =  4.00000035762787         5.00000041723251
```

Total number of zeros inside this box   =   4

Number of boxes containing at most   5 zeros =   1

These boxes are given by:

```
1) LV = -2.00000016391277        -2.00000019371510
   H  =  4.00000035762787         5.00000041723251
   Total number of zeros inside this box =    4
```

Final approximations for the zeros and verification:

1) Number of mutually distinct zeros   =   4

```
   z    = ( -1.84423395326221    ,-0.729696337329436E-29 )
   f(z) = ( 0.222044604925031E-15, 0.297690930716218E-28 )
   multiplicity =    1

   z    = ( 0.530894930292930    ,  1.33179187675112    )
   f(z) = ( 0.888178419700125E-15, 0.222044604925031E-14 )
   multiplicity =    1

   z    = ( 0.530894930292930    , -1.33179187675112    )
   f(z) = (-0.266453525910038E-14,-0.444089209850063E-15 )
   multiplicity =    1
```

```
z    = ( 0.277555756299546E-16, 0.732694008769276E-26 )
f(z) = (  0.00000000000000    , 0.366347004384638E-25 )
multiplicity =      1
```

If we set M = 2, then ZEAL outputs the following.

```
This is ZEAL.

Input:

LV    =    -2.00000000000000    -2.00000000000000
H     =     4.00000000000000     5.00000000000000

M     =    2
ICON  =    3

FILES =    T

Results:

The following box has been considered:

 LV =  -2.00000016391277      -2.00000019371510
 H  =   4.00000035762787       5.00000041723251

Total number of zeros inside this box        =     4

Number of boxes containing at most    2 zeros =    3

These boxes are given by:

  1)  LV =  -2.00000016391277       0.500000014901161
      H  =   4.00000035762787       2.50000020861626
      Total number of zeros inside this box =     1

  2)  LV =  -2.00000016391277      -2.00000019371510
      H  =   2.00000017881393       2.50000020861626
      Total number of zeros inside this box =     2

  3)  LV =  0.149011611938477E-07  -2.00000019371510
      H  =   2.00000017881393       2.50000020861626
      Total number of zeros inside this box =     1

Final approximations for the zeros and verification:
```

```
1)  Number of mutually distinct zeros    =    1

    z   = ( 0.530894930292931    ,  1.33179187675112     )
    f(z) = ( 0.888178419700125E-15,-0.177635683940025E-14 )
    multiplicity =    1

2)  Number of mutually distinct zeros    =    2

    z   = ( -1.84423395326221    ,-0.551251254781237E-21 )
    f(z) = ( 0.222044604925031E-15, 0.224891493487409E-20 )
    multiplicity =    1


    z   = (-0.501336236251204E-20,-0.135361644895767E-20 )
    f(z) = ( 0.00000000000000    ,-0.676808224478837E-20 )
    multiplicity =    1

3)  Number of mutually distinct zeros    =    1

    z   = ( 0.530894930292931    , -1.33179187675112     )
    f(z) = ( 0.888178419700125E-15,-0.444089209850063E-15 )
    multiplicity =    1
```

Finally, suppose that we want ZEAL to compute only two zeros. We set
ICON=4 and NR=2.

This is ZEAL.

Input:

```
LV    =    -2.00000000000000        -2.00000000000000
H     =     4.00000000000000         5.00000000000000

M     =    2
NR    =    2
ICON  =    4

FILES =    T
```

Results:

The following box has been considered:

```
LV =  -2.00000016391277        -2.00000019371510
H  =   4.00000035762787         5.00000041723251
```

```
Total number of zeros inside this box          =     4

Number of boxes containing at most    2 zeros =     3

These boxes are given by:

    1)  LV =  -2.00000016391277          0.500000014901161
        H  =   4.00000035762787          2.50000020861626
        Total number of zeros inside this box =    1

    2)  LV =  -2.00000016391277         -2.00000019371510
        H  =   2.00000017881393          2.50000020861626
        Total number of zeros inside this box =    2

    3)  LV =  0.149011611938477E-07    -2.00000019371510
        H  =   2.00000017881393          2.50000020861626
        Total number of zeros inside this box =    1

Requested number of mutually distinct zeros =    2

Final approximations for the zeros and verification:

    1)  Number of mutually distinct zeros     =     1

        z    = ( 0.530894930292931    , 1.33179187675112      )
        f(z) = ( 0.888178419700125E-15,-0.177635683940025E-14 )
        multiplicity =     1

    2)  Number of mutually distinct zeros     =     2

        z    = ( -1.84423395326221     ,-0.551251254781237E-21 )
        f(z) = ( 0.222044604925031E-15, 0.224891493487409E-20 )
        multiplicity =     1
```

Example 1.5.2. Suppose that $f(z) = z^2(z-1)(z-2)(z-3)(z-4) + z\sin z$ and let W be the rectangular region determined by

LV = (/-0.5_DP,-0.5_DP/) and H = (/ 6.0_DP, 2.0_DP/).

Note that f has a double zero at the origin. We set M=5 and ICON=3.

This is ZEAL.

Input:

```
LV    =   -0.500000000000000      -0.500000000000000
H     =    6.00000000000000       2.00000000000000


M     =   5
ICON  =   3

FILES =   T
```

Results:

The following box has been considered:

```
  LV = -0.500000163912773        -0.500000193715096
  H  =   6.00000035762787         2.00000041723251
```

Total number of zeros inside this box   =   6

Number of boxes containing at most   5 zeros =   2

These boxes are given by:

```
  1)  LV = -0.500000163912773        -0.500000193715096
      H  =   3.00000017881393         2.00000041723251
      Total number of zeros inside this box =      4

  2)  LV =   2.50000001490116        -0.500000193715096
      H  =   3.00000017881393         2.00000041723251
      Total number of zeros inside this box =      2
```

Final approximations for the zeros and verification:

```
  1)  Number of mutually distinct zeros      =      3

        z    = (-0.444089209850063E-15,-0.284397851396988E-16 )
        f(z) = ( 0.491016012316152E-29, 0.631490085549722E-30 )
        multiplicity =      2

        z    = (  1.18906588973011    , 0.840925724965599E-27 )
        f(z) = (  0.00000000000000    ,-0.332912397884017E-26 )
        multiplicity =      1

        z    = (  1.72843498616506    ,-0.366031460022549E-27 )
        f(z) = (-0.222044604925031E-15,-0.174867254641548E-26 )
        multiplicity =      1
```

```
2)  Number of mutually distinct zeros     =     2

    z    = (  4.03038191606047      , 0.288920306537196E-28 )
    f(z) = ( 0.155431223447522E-13, 0.308650229769291E-26 )
    multiplicity =      1


    z    = (  3.01990732809571      , 0.185382312726938E-28 )
    f(z) = (-0.105471187339390E-14,-0.402275402644635E-27 )
    multiplicity =      1
```

Example 1.5.3. Finally, suppose that $f(z) = z^2(z-2)^2[e^{2z}\cos z + z^3 - 1 - \sin z]$ and let W be the region determined by

LV = (/-1.0_DP,-1.0_DP/) and H = (/ 4.0_DP, 2.0_DP/).

Note that f has a triple zero at the origin and a double zero at $z = 2$. We set M=5 and ICON=3.

```
This is ZEAL.

Input:

LV     =     -1.00000000000000        -1.00000000000000
H      =      4.00000000000000         2.00000000000000

M      =     5
ICON   =     3

FILES  =     T

Results:

The following box has been considered:

LV =   -1.00000016391277        -1.00000019371510
 H =    4.00000035762787         2.00000041723251

Total number of zeros inside this box          =     8

Number of boxes containing at most     5 zeros =     2

These boxes are given by:

1)  LV =   -1.00000016391277        -1.00000019371510
     H =    2.00000017881393         2.00000041723251
```

```
            Total number of zeros inside this box =     5

  2)  LV =   1.00000001490116        -1.00000019371510
      H  =   2.00000017881393         2.00000041723251
      Total number of zeros inside this box =     3
```

Final approximations for the zeros and verification:

```
  1)  Number of mutually distinct zeros      =     3

      z   = (-0.460714119728971     , 0.625427769347768     )
      f(z) = (-0.125408855493498E-14,-0.242634956938915E-14 )
      multiplicity =      1


      z   = (-0.555111512312578E-16, 0.780756160460674E-15 )
      f(z) = ( 0.270707981407843E-45,-0.189411036718915E-44 )
      multiplicity =      3


      z   = (-0.460714119728972     ,-0.625427769347767     )
      f(z) = ( 0.420110659916658E-14, 0.555338820267851E-14 )
      multiplicity =      1

  2)  Number of mutually distinct zeros      =     2

      z   = ( 2.00000000000000     , 0.105626554996077E-14 )
      f(z) = (-0.253754972900138E-27,-0.312032161709746E-27 )
      multiplicity =      2


      z   = ( 1.66468286974552     ,-0.528536806498078E-28 )
      f(z) = ( 0.276741773088933E-15, 0.405540193714869E-27 )
      multiplicity =      1
```

1.5.5 Concluding remarks

We have applied our package to various analytic functions and rectangular regions and we have found that it behaves predictably and accurately. ZEAL calculates the total number of zeros that lie inside the given box and then computes approximations for these zeros, together with their respective multiplicities. Our package does not require initial approximations for the zeros.

The user will appreciate the flexibility offered by the input parameter ICON. If nothing is known about the zeros that lie inside the given box, one may call ZEAL with ICON = 1 to obtain the total number of zeros. Then one may proceed with ICON = 3 to compute approximations for all these zeros, or, if less than the total number of zeros are required, with ICON = 4 and

NR equal to the requested number of zeros. If only a set of boxes is required, each of which contains less than M zeros (counting multiplicities), then one may set ICON = 2.

1.6 A derivative-free approach

The results presented in the previous sections are based on the form (1.12), which involves the logarithmic derivative f'/f. Instead, let us consider the symmetric bilinear form

$$\langle \cdot, \cdot \rangle_\star \ : \ \mathcal{P} \times \mathcal{P} \to \mathbb{C}$$

defined as

$$\langle \phi, \psi \rangle_\star := \frac{1}{2\pi i} \int_\gamma \phi(z)\psi(z) \frac{1}{f(z)} \, dz \qquad (1.26)$$

for any two polynomials $\phi, \psi \in \mathcal{P}$. Again, this form can be evaluated via numerical integration along γ and in what follows we will assume that all the "inner products" $\langle \phi, \psi \rangle_\star$ that are needed have been calculated.

We will show that essentially the same results can be obtained with the form $\langle \cdot, \cdot \rangle_\star$ as previously with $\langle \cdot, \cdot \rangle$. Note that the derivative f' is no longer needed. Of course, in this new approach not the mutually distinct zeros but rather the unknowns $Z_1, \ldots, Z_N$ (introduced by Delves and Lyness, cf. Section 1.1) are calculated and the multiplicities cannot be computed explicitely. But apart from this, the approach has the same advantages as the algorithm (which we will henceforth call 'our Algorithm') that we have presented in Section 1.3. In particular, it is self-starting in the sense that it does not require initial approximations for the zeros.

As by assumption the derivative f' is not available to us, we cannot obtain the value of N by evaluating the integral in the right-hand side of (1.3) numerically. Instead one can use the principle of the argument, cf. our discussion in Subsection 1.1.1.

The integrand that appears in the right-hand side of (1.26) has a pole at every zero of f that lies in the interior of γ and the order of the pole is equal to the multiplicity of the zero. Therefore, the residue theorem implies that $\langle \phi, \psi \rangle_\star$ is equal to the sum of the residues of the function $\phi\psi/f$ at these poles. The following result can easily be verified.

Proposition 1.6.1. *Suppose that all the N zeros $Z_1, \ldots, Z_N$ of f that lie inside γ are simple. Then*

$$\langle \phi, \psi \rangle_\star = \sum_{k=1}^{N} \frac{\phi(Z_k)\psi(Z_k)}{f'(Z_k)}.$$

In general, if f has multiple zeros, then an elegant expression for $\langle \phi, \psi \rangle_\star$ written as a sum is much more difficult to obtain. Fortunately, it is not necessary to have such an expression available. The proofs of Theorems 1.2.1 and 1.2.2 depend completely on the details of the way in which $\langle \phi, \psi \rangle$ can be written as a sum, cf. Equation (1.12). However, as we will see, the corresponding theorems can also be proved in a different way.

Define $s_p^\star := \langle 1, z^p \rangle_\star$ for $p = 0, 1, 2, \ldots$ and let $H_k^\star$ be the $k \times k$ Hankel matrix

$$H_k^\star := \left[s_{p+q}^\star \right]_{p,q=0}^{k-1}$$

for $k = 1, 2, \ldots$. The formal orthogonal polynomials associated with $\langle \cdot, \cdot \rangle_\star$ can be defined as before. The coefficients of regular FOPs can be computed by solving a Yule-Walker system, cf. Equation (1.14). Also, $t \geq 1$ is a regular index if and only if the matrix $H_t^\star$ is nonsingular.

The residue theorem immediately implies that the polynomial

$$P_N(z) = \prod_{k=1}^{N} (z - Z_k)$$

satisfies

$$\langle z^p, P_N(z) \rangle_\star = 0, \qquad p = 0, 1, 2, \ldots. \tag{1.27}$$

In this sense, the polynomial $P_N(z)$ behaves with respect to the form $\langle \cdot, \cdot \rangle_\star$ in the same way as the polynomial $\varphi_n(z)$ behaves with respect to $\langle \cdot, \cdot \rangle$, cf. Equations (1.16) and (1.17). We will prove that N is the largest regular index for $\langle \cdot, \cdot \rangle_\star$. This will enable us to compute the zeros of the regular FOP $P_N(z)$, i.e., the zeros $Z_1, \ldots, Z_N$, in essentially the same way as our Algorithm applied to the form $\langle \cdot, \cdot \rangle$ computes the zeros of $\varphi_n(z)$, i.e., the mutually distinct zeros $z_1, \ldots, z_n$.

The following lemma will play an important role. Define the set $\mathcal{I}$ as follows:

$$\mathcal{I} := \{ \phi \in \mathcal{P} : \langle z^p, \phi(z) \rangle_\star = 0 \quad \text{for } p = 0, 1, 2, \ldots \}.$$

Lemma 1.6.1. *The set $\mathcal{I}$ is equal to the ideal generated by the polynomial P_N. In other words,*

$$\mathcal{I} = \{ \phi \in \mathcal{P} : \exists \alpha \in \mathcal{P} : \phi = \alpha P_N \}.$$

Proof. Suppose that $a \in \mathbb{C}$ lies in the interior of γ. Let the function $g : W \to \mathbb{C}$ be meromorphic and suppose that g has neither zeros nor poles on γ. Then the coefficient of $(z - a)^{-p-1}$ in the Laurent expansion of g at the point a is given by the integral

$$\frac{1}{2\pi i} \int_\gamma (z - a)^p \, g(z) \, dz$$

for $p = 0, 1, 2, \ldots$. Let $\phi \in \mathcal{I}$. Then

$$\langle (z - Z_k)^p, \phi(z) \rangle_\star = \frac{1}{2\pi i} \int_\gamma (z - Z_k)^p \frac{\phi(z)}{f(z)} \, dz = 0$$

for $k = 1, \ldots, N$ and $p = 0, 1, 2, \ldots$ and thus the function ϕ/f has a removable singularity at the points $Z_1, \ldots, Z_N$. Thus ϕ has to be a multiple of P_N. This proves the lemma. □

Theorem 1.6.1. *The matrix $H_N^\star$ is nonsingular.*

Proof. We will prove that P_N is the only monic polynomial of degree N that is orthogonal to all polynomials of lower degree. Suppose that Q_N is another such polynomial. Then $P_N - Q_N$ is of degree at most $N - 1$ and hence $\langle P_N - Q_N, Q_N \rangle_\star = 0$. Equation (1.27) then implies that $\langle Q_N, Q_N \rangle_\star = 0$. Thus Q_N is not only orthogonal to all polynomials of degree $\leq N - 1$ but also to all polynomials of degree N. The polynomial $zP_N - zQ_N$ has degree $\leq N$ and therefore $\langle zP_N - zQ_N, Q_N \rangle_\star = 0$. As $\langle zP_N, Q_N \rangle_\star = \langle P_N, zQ_N \rangle_\star = 0$, it follows that $\langle zQ_N, Q_N \rangle_\star = 0$. Thus Q_N is also orthogonal to all polynomials of degree $N + 1$. By continuing this way, one can prove that Q_N is orthogonal to *all* polynomials, $Q_N \in \mathcal{I}$. As Q_N is a monic polynomial of degree N, the previous lemma then implies that $Q_N = P_N$. Thus there is only one monic polynomial of degree N that is orthogonal to all polynomials of lower degree. This implies that the matrix $H_N^\star$ is nonsingular. □

Theorem 1.6.2. *The matrix $H_{N+k}^\star$ is singular for $k = 1, 2, \ldots$.*

Proof. Instead of the basis of the monomials $\{z^p\}_{p \geq 0}$ we consider the basis $\{\psi_p(z)\}_{p \geq 0}$ where $\psi_p(z) := z^p$ for $p = 0, 1, \ldots, N - 1$ and $\psi_{N+p}(z) := z^p P_N(z)$ for $p = 0, 1, 2, \ldots$. Let

$$F_l^\star := \left[\langle \psi_p, \psi_q \rangle_\star \right]_{p,q=0}^{l-1}$$

be the corresponding $l \times l$ Gram matrix for $l = 1, 2, \ldots$. Equation (1.27) then implies that $\det F_{N+k}^\star = 0$ for $k = 1, 2, \ldots$. One can easily verify that $\det F_l^\star = \det H_l^\star$ for $l = 1, 2, \ldots$. This proves the theorem. □

We have now identified $P_N(z)$ as the regular FOP of degree N and we have shown that regular FOPs of degree larger than N do not exist. Note that $s_0^\star$ is equal to the sum of the residues of $1/f$ at the points $Z_1, \ldots, Z_N$ and hence it is not necessarily different from zero. Therefore, the regular FOP of degree 1 with respect to the form $\langle \cdot, \cdot \rangle_\star$ doesn't always exist, in contrast to $\langle \cdot, \cdot \rangle$.

The zero/eigenvalue properties discussed in Section 1.3 hold not only for $\langle \cdot, \cdot \rangle$ but for every symmetric bilinear form. The zeros $Z_1, \ldots, Z_N$ can therefore also be calculated by solving a generalized eigenvalue problem. The following result can be proved in the same way as Theorem 1.2.2. Let $H_k^{\star<}$ be the Hankel matrix

$$H_k^{\star<} := \begin{bmatrix} s_1^\star & s_2^\star & \cdots & & s_k^\star \\ s_2^\star & & & \cdot^{\cdot^{\cdot}} & \vdots \\ \vdots & & \cdot^{\cdot^{\cdot}} & & \vdots \\ s_k^\star & \cdots & \cdots & & s_{2k-1}^\star \end{bmatrix}$$

for $k = 1, 2, \ldots$.

Theorem 1.6.3. *The eigenvalues of the pencil* $H_N^{\star<} - \lambda H_N^\star$ *are given by* $Z_1, \ldots, Z_N$.

Below we will compare the accuracy obtained via Theorem 1.2.2 to the accuracy obtained via Theorem 1.6.3.

The zeros $Z_1, \ldots, Z_N$ can also be computed by applying our Algorithm to the form $\langle \cdot, \cdot \rangle_\star$. Let $\{\varphi_t^\star\}_{t \geq 0}$ denote the FOPs associated with $\langle \cdot, \cdot \rangle_\star$. Define the matrices $G_k^\star$ and $G_k^{\star z}$ as

$$G_k^\star := \left[\langle \varphi_r^\star, \varphi_s^\star \rangle \right]_{r,s=0}^{k-1} \quad \text{and} \quad G_k^{\star z} := \left[\langle \varphi_r^\star, z\varphi_s^\star \rangle \right]_{r,s=0}^{k-1}$$

for $k = 1, 2, \ldots$. The following results can be proved in the same way as Theorem 1.3.1 and Corollary 1.3.1.

Theorem 1.6.4. *Let* $t \geq 1$ *be a regular index for* $\langle \cdot, \cdot \rangle_\star$ *and let* $z_{t,1}^\star, \ldots, z_{t,t}^\star$ *be the zeros of the regular FOP* $\varphi_t^\star$. *Then the eigenvalues of the pencil* $G_t^{\star z} - \lambda G_t^\star$ *are given by* $z_{t,1}^\star, \ldots, z_{t,t}^\star$.

Corollary 1.6.1. *The eigenvalues of* $G_N^{\star z} - \lambda G_N^\star$ *are given by* $Z_1, \ldots, Z_N$.

If instead of N only an upper bound for N is available, then the value of N can be computed via the stopping criterion of our Algorithm.

We will now discuss a few numerical examples. The computations have been done via Matlab 5 (with floating point relative accuracy $\approx 2.2204 10^{-16}$). The integration algorithm is the same as the one discussed at the beginning of Section 1.4.

Example 1.6.1. Let $f(z) = e^{3z} - 2z \cos z - 1$. Suppose that γ is the circle $\gamma = \{ z \in \mathbb{C} : |z| = 4 \}$. Then $N = 6$. Let us try an approach based on ordinary moments. Table 1.1 contains approximations for $s_p = \langle 1, z^p \rangle$ and $s_p^\star = \langle 1, z^p \rangle_\star$ for $p = 0, 1, \ldots, 11$. Note that in both cases the order of magnitude changes as p increases. The computed approximations for the zeros $Z_1, \ldots, Z_N$ obtained via Theorem 1.2.2 and 1.6.3 are shown in Table 1.2 and 1.3, respectively. The digits that are not correct are underlined. Observe

p	s_p	$s_p^\star$
0	6.0	$-7.5\,10^{-2}$
1	2.0	$2.8\,10^{-1}$
2	$-1.4\,10^1$	$-9.1\,10^{-1}$
3	$-8.5\,10^1$	2.1
4	$-3.0\,10^1$	-2.0
5	$7.0\,10^2$	2.4
6	$2.5\,10^3$	$-2.3\,10^1$
7	$-1.0\,10^3$	1.5 101
8	$-3.1\,10^4$	$9.9\,10^1$
9	$-7.6\,10^4$	$4.6\,10^2$
10	$1.3\,10^5$	$-4.8\,10^2$
11	$1.2\,10^6$	$-5.3\,10^3$

Table 1.1. Ordinary moments s_p and $s_p^\star$

$-2.\underline{186079491175828}\,10^{-13}$	$-$	$i\,\underline{1.727623083122153}\,10^{-12}$
$5.308949302929\underline{420}\,10^{-1}$	$+$	$i\,1.331791876750\underline{615}$
$5.308949302928\underline{376}\,10^{-1}$	$-$	$i\,1.331791876751\underline{221}$
$-1.84423395326\underline{2199}$	$-$	$i\,4.\underline{204494152042317}\,10^{-14}$
$1.4146071776581\underline{90}$	$+$	$i\,3.047722062627\underline{169}$
$1.4146071776581\underline{85}$	$-$	$i\,3.047722062627173$

Table 1.2. Approximations for the zeros obtained via the ordinary moments s_p.

$5.\underline{879198486593449}\,10^{-14}$	$+$	$i\,\underline{1.896836726398249}\,10^{-14}$
$5.308949302929\underline{366}\,10^{-1}$	$+$	$i\,1.331791876751\underline{066}$
$5.308949302929\underline{205}\,10^{-1}$	$-$	$i\,1.331791876751\underline{080}$
-1.844233953262213	$-$	$i\,\underline{1.231196347425826}\,10^{-15}$
$1.41460717765818\underline{1}$	$+$	$i\,3.047722062627\underline{166}$
$1.41460717765818\underline{0}$	$-$	$i\,3.047722062627\underline{167}$

Table 1.3. Approximations for the zeros obtained via the ordinary moments $s_p^\star$.

that the approximations for the zeros are very accurate. Using ordinary moments has the advantage that only $2N$ integrals have to be calculated and hence, compared to our Algorithm, the arithmetic cost is rather limited. Also, a significant part of the computation required for each integrand is the same for all of the integrands (namely, the computation of f'/f or $1/f$). By programming the quadrature algorithm in such a way that it is able to integrate a *vector* of similar integrals, these common calculations need be done only once for each integrand evaluation point. However, as the following example shows, ordinary moments do not always lead to such accurate results.

Example 1.6.2. The Wilkinson polynomial and also functions that have clusters of zeros are typical, although somewhat extreme, examples where an approach based on ordinary moments is likely to fail. The following function is another example. Suppose that $f(z) = J_0(z)$, the Bessel function of the first kind and of order zero. It is known that this function has only positive real zeros and that all these zeros are simple (see, e.g., Watson [123]). In that sense it is related to the Wilkinson polynomial. Suppose that $\gamma = \{ z \in \mathbb{C} : |z - 15| = 14.5 \}$. Then $N = 9$. Table 1.4 gives for each zero the number of correct significant digits obtained via the ordinary moments s_p^* (Theorem 1.6.3), our Algorithm applied to the form $\langle \cdot, \cdot \rangle_*$, the ordinary moments s_p (Theorem 1.2.2) and our Algorithm applied to the form $\langle \cdot, \cdot \rangle$.

exact zeros	s_p^*	$\langle \cdot, \cdot \rangle_*$	s_p	$\langle \cdot, \cdot \rangle$
2.404825557695773	5	12	6	12
5.520078110286311	2	11	4	10
8.653727912911013	2	10	4	9
11.79153443901428	3	11	5	9
14.93091770848778	3	11	5	9
18.07106396791092	3	11	4	10
21.21163662987926	4	11	5	11
24.35247153074930	5	11	6	11
27.49347913204025	7	11	7	12

Table 1.4. The number of correct significant digits in case $f(z) = J_0(z)$.

Observe that in both cases the approximations obtained via our Algorithm are more accurate than the approximations obtained via ordinary moments. Of course, there is clearly a trade-off between obtained accuracy and cost. We advise the reader to start with the cheapest approach, i.e., the approach based on the ordinary moments s_p^*. If the computed approximations for the zeros are not sufficiently accurate to be refined via an iterative method (one that doesn't need the derivative, of course), then one can apply our Algorithm to the form $\langle \cdot, \cdot \rangle_*$ or switch to one of the approaches that use both f and f'.

Example 1.6.3. Let us illustrate how the stopping criterion of our Algorithm can be used to determine the value of N in case only an upper bound for N is known. Consider again the function $f(z) = e^{3z} - 2z \cos z - 1$ and suppose that γ is the circle $\gamma = \{ z \in \mathbb{C} : |z| = 5 \}$. Then $N = 7$. Let us assume that only the upper bound 20 is known. Our algorithm defines the FOP φ_1^* as an inner polynomial and φ_2^* as a regular FOP. At this point the algorithm asks itself whether N is equal to two. It computes $|\langle \varphi_2^*, \varphi_2^* \rangle_*|$. To take into account the accuracy lost during the evaluation of the quadrature formula, this quantity is scaled in a certain way, cf. Section 1.3. The resulting floating point number is given by

1.998545018990362,

which is certainly not "sufficiently small" (we use 10^{-8} as a threshold) and hence the algorithm continues. It defines $\varphi_3^\star$ as an inner polynomial and $\varphi_4^\star$ as a regular FOP. Then it checks if N is equal to four. It compares

$$1.981687581683116$$

to 10^{-8} and continues. The polynomial $\varphi_5^\star$ is defined as a regular FOP. The algorithm again decides to continue and defines $\varphi_6^\star$ as a regular FOP. The corresponding floating point number is given by

$$0.3794164188056766$$

and the algorithm continues. It defines $\varphi_7^\star$ as a regular FOP. We have now reached the actual value of N. The scaled counterparts of the inner products that correspond to the sequence (1.23) are given by

$$9.190814944765118 \; 10^{-16}$$
$$1.799485800789563 \; 10^{-15}$$
$$4.008700446099430 \; 10^{-15}$$
$$5.548436809880727 \; 10^{-15}$$
$$6.603511781861113 \; 10^{-15}$$
$$5.538342691314587 \; 10^{-15}$$
$$3.494634208761963 \; 10^{-15}$$
$$4.116380174988637 \; 10^{-16}$$
$$5.379567405602837 \; 10^{-15}$$
$$8.806423950940129 \; 10^{-15}$$
$$8.912210606016112 \; 10^{-15}$$
$$5.866528582137192 \; 10^{-15}$$
$$1.127215921207880 \; 10^{-15}$$

and hence the algorithm decides that N is equal to seven and it stops. The computed approximations for the zeros are given by

$$-2.212860324230451 \; 10^{-11} \quad + \quad i\,5.610894531592185 \; 10^{-12}$$
$$5.308949303037738 \; 10^{-1} \quad + \quad i\,1.331791876751059$$
$$5.308949303027991 \; 10^{-1} \quad - \quad i\,1.331791876755293$$
$$-1.844233953258748 \quad - \quad i\,1.244550552500599 \; 10^{-12}$$
$$1.414607177657119 \quad + \quad i\,3.047722062626751$$
$$1.414607177657241 \quad - \quad i\,3.047722062626826$$
$$-4.603562881675490 \quad + \quad i\,3.443757237606488 \; 10^{-14}$$

The correct significant digits are underlined. Let us now compare this with the approach based on ordinary moments. The following theorem generalizes Theorem 1.6.3.

Theorem 1.6.5. *Let t be an integer $\geq N$. The eigenvalues of the pencil $H_t^{\star <} - \lambda H_t^\star$ are given by the zeros $Z_1, \ldots, Z_N$ and $t - N$ eigenvalues that may assume arbitrary values.*

Proof. Instead of the basis of the monomials $\{z^p\}_{p \geq 0}$ we consider again the basis $\{\psi_p(z)\}_{p \geq 0}$ where $\psi_p(z) := z^p$ for $p = 0, 1, \ldots, N-1$ and $\psi_{N+p}(z) := z^p P_N(z)$ for $p = 0, 1, 2, \ldots$, cf. the proof of Theorem 1.6.2. Define

$$F_t^{\star <} := \left[\langle \psi_p, z \psi_q \rangle_* \right]_{p,q=0}^{t-1} \quad \text{and} \quad F_t^{\star} := \left[\langle \psi_p, \psi_q \rangle_* \right]_{p,q=0}^{t-1}.$$

Then one can easily show that the generalized eigenvalue problem $H_t^{\star <} x = \lambda H_t^{\star} x$ is equivalent to the problem $F_t^{\star <} y = \lambda F_t^{\star} y$. Here $y := U_t^{-1} x$ where U_t denotes the unit upper triangular matrix that contains the coefficients (in the standard monomial basis) of the polynomials $\psi_0(z), \psi_1(z), \ldots, \psi_{t-1}(z)$. Equation (1.27) then implies that

$$F_t^{\star <} = \begin{bmatrix} H_N^{\star <} & 0 \\ 0 & 0 \end{bmatrix} \quad \text{and} \quad F_t^{\star} = \begin{bmatrix} H_N^{\star} & 0 \\ 0 & 0 \end{bmatrix}.$$

This proves the theorem. □

Each of these indeterminate generalized eigenvalues corresponds to two corresponding zeros on the diagonals of the generalized Schur decomposition of the Hankel matrices $H_t^{\star <}$ and $H_t^{\star}$. When actually calculated, these diagonal entries are different from zero because of roundoff errors and Matlab returns their quotient as an eigenvalue. Thus, by solving the 20×20 generalized eigenvalue problem $H_{20}^{\star <} - \lambda H_{20}^{\star}$ we obtain approximations for the seven zeros $Z_1, \ldots, Z_N$ and 13 spurious eigenvalues. The latter can be detected by evaluating f at the computed eigenvalues and also by taking into account that the computed approximations for the zeros are likely to lie inside γ or at least quite close to it. The approximations for the zeros obtained in this way are 1 to 3 digits less accurate than the approximations obtained via our Algorithm. By solving the 7×7 generalized eigenvalue problem, one obtains approximations that are about as accurate as those computed by our Algorithm.

Example 1.6.4. Let us consider a function that has multiple zeros. Suppose that

$$f(z) = z^2 (z-2)^2 [e^{2z} \cos z + z^3 - 1 - \sin z]$$

and let $\gamma = \{z \in \mathbb{C} : |z| = 3\}$. Note that f has a triple zero at the origin and a double zero at $z = 2$. The total number of zeros of f that lie inside γ is equal to eight, $N = 8$. By using the ordinary moments s_p^* we obtain the following approximations for the zeros:

$1.183531315599526 \ 10^{-4}$	$-$	$i \ 8.840648137844101 \ 10^{-7}$
$-5.994094794302300 \ 10^{-5}$	$-$	$i \ 1.020522445441414 \ 10^{-4}$
$-5.841218335364184 \ 10^{-5}$	$+$	$i \ 1.029363093460768 \ 10^{-4}$
2.000000113260292	$+$	$i \ 9.253727402009306 \ 10^{-7}$
1.999999886743732	$-$	$i \ 9.253736788382785 \ 10^{-7}$
$-4.607141197285995 \ 10^{-1}$	$+$	$i \ 6.254277693471380 \ 10^{-1}$
$-4.607141197287246 \ 10^{-1}$	$-$	$i \ 6.254277693472881 \ 10^{-1}$
1.664682869740608	$+$	$i \ 1.093307455221265 \ 10^{-12}$

We have underlined the correct significant digits. Our Algorithm gives comparable results. Note how the obtained accuracy diminishes as the multiplicity of the zero increases.

2. Clusters of zeros of analytic functions

In the previous chapter we have presented an accurate algorithm, based on the theory of formal orthogonal polynomials, for computing zeros of analytic functions. More specifically, given an analytic function f and a Jordan curve γ that does not pass through any zero of f, we have considered the problem of computing *all* the zeros $z_1, \ldots, z_n$ of f that lie inside γ, together with their respective multiplicities $\nu_1, \ldots, \nu_n$. Our principal means of obtaining information about the location of these zeros has been the symmetric bilinear form $\langle \cdot, \cdot \rangle$, cf. Equation (1.12). This form can be evaluated via numerical integration along γ.

This chapter continues the previous chapter. If f has one or several clusters of zeros, then the mapping from the ordinary moments associated with $\langle \cdot, \cdot \rangle$ to the zeros and their respective multiplicities is very ill-conditioned. We will show that the algorithm that we have presented in Chapter 1 can be used to calculate the centre of a cluster and its size, i.e., the arithmetic mean of the zeros that form a certain cluster and the total number of zeros in this cluster, respectively. This information enables one to zoom into a certain cluster: its zeros can be calculated separately from the other zeros of f. By shifting the origin in the complex plane to the centre of a certain cluster, its zeros become better relatively separated, which is appropriate in floating point arithmetic and reduces the ill-conditioning.

In this chapter we will also attack our problem of computing all the zeros of f that lie inside γ in an entirely different way, based on rational interpolation at roots of unity. We will show how the new approach complements the previous one and how it can be used effectively in case γ is the unit circle.

Note 2.0.1. Specifically for clusters of polynomial zeros, let us mention that Hribernig and Stetter [70] worked on detection and validation of clusters of zeros whereas Kirrinnis [84] studied Newton's iteration towards a cluster.

2.1 How to obtain the centre of a cluster and its weight

Suppose that the zeros of f that lie inside γ can be grouped into m clusters. Let $I_1, \ldots, I_m$ be index sets that define these clusters, and let

$$\mu_j := \sum_{k \in I_j} \nu_k \quad \text{and} \quad c_j := \frac{1}{\mu_j} \sum_{k \in I_j} \nu_k z_k$$

for $j = 1, \ldots, m$. In other words, μ_j is equal to the total number of zeros that form cluster j (its "weight") whereas c_j is equal to the arithmetic mean of the zeros in cluster j (its "centre of gravity"). We assume that the centres $c_1, \ldots, c_m$ are mutually distinct. For $k = 1, \ldots, n$ we also define $\zeta_k := z_k - c_j$ if $k \in I_j$. From the definition of μ_j and c_j it follows that

$$\sum_{k \in I_j} \nu_k \zeta_k = 0, \qquad j = 1, \ldots, m.$$

Define the symmetric bilinear form $\langle \cdot, \cdot \rangle_m$ by

$$\langle \phi, \psi \rangle_m := \sum_{j=1}^{m} \mu_j \phi(c_j) \psi(c_j)$$

for any two polynomials $\phi, \psi \in \mathcal{P}$. This form is related to the form $\langle \cdot, \cdot \rangle$ in an obvious way: instead of the zeros $z_1, \ldots, z_n$ and their multiplicities $\nu_1, \ldots, \nu_n$, we now use the centres of gravity $c_1, \ldots, c_m$ and the weights $\mu_1, \ldots, \mu_m$ of the clusters. Let

$$\delta := \max_{1 \le k \le n} |\zeta_k|.$$

The following theorem tells us that $\langle \cdot, \cdot \rangle_m$ approximates $\langle \cdot, \cdot \rangle$ (and vice versa).

Theorem 2.1.1. Let $\phi, \psi \in \mathcal{P}$. Then $\langle \phi, \psi \rangle = \langle \phi, \psi \rangle_m + \mathcal{O}(\delta^2)$, $\delta \to 0$.

Proof. The following holds:

$$
\begin{aligned}
\langle \phi, \psi \rangle &= \sum_{k=1}^{n} \nu_k \phi(z_k) \psi(z_k) \\
&= \sum_{j=1}^{m} \sum_{k \in I_j} \nu_k \phi(c_j + \zeta_k) \psi(c_j + \zeta_k) \\
&= \sum_{j=1}^{m} \sum_{k \in I_j} \nu_k \left(\phi(c_j) \psi(c_j) + \zeta_k [\phi(z) \psi(z)]'_{z=c_j} + \mathcal{O}(\zeta_k^2), \ \zeta_k \to 0 \right) \\
&= \sum_{j=1}^{m} \mu_j \phi(c_j) \psi(c_j) + \sum_{j=1}^{m} \underbrace{\left(\sum_{k \in I_j} \nu_k \zeta_k \right)}_{=0} [\phi(z) \psi(z)]'_{z=c_j} + \sum_{k=1}^{n} \mathcal{O}(\zeta_k^2), \ \zeta_k \to 0 \\
&= \sum_{j=1}^{m} \mu_j \phi(c_j) \psi(c_j) + \sum_{k=1}^{n} \mathcal{O}(\zeta_k^2), \ \zeta_k \to 0.
\end{aligned}
$$

This proves the theorem. $\qquad\qquad\qquad\qquad\qquad\qquad\qquad\qquad\qquad\qquad\qquad\qquad$ □

Define the ordinary moments associated with $\langle \cdot, \cdot \rangle_m$ as

$$s_p^{(m)} := \langle 1, z^p \rangle_m$$

for $p = 0, 1, 2, \ldots$. Observe that $s_0^{(m)} = s_0$ whereas $s_1^{(m)} = s_1$. Define the vectors $\mathbf{s}, \mathbf{s}^{(m)} \in \mathbb{C}^{2N-1}$ as

$$\mathbf{s} := \begin{bmatrix} s_0 \\ s_1 \\ \vdots \\ s_{2N-2} \end{bmatrix} \quad \text{and} \quad \mathbf{s}^{(m)} := \begin{bmatrix} s_0^{(m)} \\ s_1^{(m)} \\ \vdots \\ s_{2N-2}^{(m)} \end{bmatrix}.$$

The entries of $\mathbf{s}$ determine the Hankel matrix H_N. The previous theorem implies that

$$\frac{\|\mathbf{s} - \mathbf{s}^{(m)}\|_2}{\|\mathbf{s}\|_2} = \mathcal{O}(\delta^2), \; \delta \to 0. \tag{2.1}$$

Let $H_k^{(m)}$ be the $k \times k$ Hankel matrix

$$H_k^{(m)} := \left[s_{p+q}^{(m)} \right]_{p,q=0}^{k-1}$$

for $k = 1, 2, \ldots$.

Corollary 2.1.1. *Let $k \geq 1$. Then* $\det H_k = \det H_k^{(m)} + \mathcal{O}(\delta^2), \; \delta \to 0$.

Proof. Let k be a positive integer. Then the previous theorem implies that

$$H_k = \left[s_{p+q} \right]_{p,q=0}^{k-1} = \left[s_{p+q}^{(m)} + \mathcal{O}(\delta^2), \; \delta \to 0 \right]_{p,q=0}^{k-1}.$$

The result follows by expanding the determinant of the matrix in the right-hand side.   □

Corollary 2.1.2. *The matrix H_m is nonsingular if $\delta \to 0$. Let $t > m$. Then* $\det H_t = \mathcal{O}(\delta^2), \; \delta \to 0$.

Proof. This follows from the previous corollary and the fact that $H_m^{(m)}$ is non-singular whereas $H_t^{(m)}$ is singular for all integers $t > m$ (cf. Theorem 1.2.1). □

The following theorem should be compared with Theorem 1.3.2.

Theorem 2.1.2. *Let t be an integer $\geq m$. Then $\varphi_t(c_j) = \mathcal{O}(\delta^2), \; \delta \to 0$ for $j = 1, \ldots, m$. Also $\langle z^p, \varphi_t(z) \rangle = \mathcal{O}(\delta^2), \; \delta \to 0$ for all $p \geq t$.*

Proof. Let $t \geq m$. If t is a regular index, then

$$\langle z^p, \varphi_t(z) \rangle = 0, \qquad p = 0, 1, \ldots, t-1,$$

else

$$\langle z^p, \varphi_t(z) \rangle = 0, \qquad p = 0, 1, \ldots, r-1$$

where r is the largest regular index less than t. Corollary 2.1.2 implies that $r \geq m$, and thus we may conclude that

$$\langle z^p, \varphi_t(z) \rangle = 0, \qquad p = 0, 1, \ldots, m-1.$$

Theorem 2.1.1 then implies that

$$\langle z^p, \varphi_t(z) \rangle_m = \mathcal{O}(\delta^2), \ \delta \to 0, \qquad p = 0, 1, \ldots, m-1.$$

In matrix notation this can be written as

$$
\begin{bmatrix}
1 & \cdots & 1 \\
c_1 & \cdots & c_m \\
\vdots & & \vdots \\
c_1^{m-1} & \cdots & c_m^{m-1}
\end{bmatrix}
\begin{bmatrix}
\mu_1 & & & \\
& \mu_2 & & \\
& & \ddots & \\
& & & \mu_m
\end{bmatrix}
\begin{bmatrix}
\varphi_t(c_1) \\
\varphi_t(c_2) \\
\vdots \\
\varphi_t(c_m)
\end{bmatrix}
= \mathcal{O}(\delta^2), \ \delta \to 0.
$$

The right-hand side represents a vector in $\mathbb{C}^m$ whose entries are $\mathcal{O}(\delta^2)$, $\delta \to 0$. As the centres $c_1, \ldots, c_m$ are assumed to be mutually distinct and the weights $\mu_1, \ldots, \mu_m$ are different from zero, it follows that

$$\varphi_t(c_j) = \mathcal{O}(\delta^2), \ \delta \to 0, \qquad j = 1, \ldots, m.$$

Theorem 2.1.1 then immediately implies that

$$\langle z^p, \varphi_t(z) \rangle = \mathcal{O}(\delta^2), \ \delta \to 0$$

for all $p \geq t$. $\qquad \square$

In other words, unless the FOP $\varphi_t(z)$ has a very flat shape near its zeros, we are likely to find good approximations for the centres $c_1, \ldots, c_m$ among the zeros of $\varphi_t(z)$ for all $t \geq m$. Note that

$$
\begin{bmatrix}
1 & \cdots & 1 \\
c_1 & \cdots & c_m \\
\vdots & & \vdots \\
c_1^{m-1} & \cdots & c_m^{m-1}
\end{bmatrix}
\begin{bmatrix}
\mu_1 \\
\mu_2 \\
\vdots \\
\mu_m
\end{bmatrix}
=
\begin{bmatrix}
s_0 \\
s_1 \\
\vdots \\
s_{m-1}
\end{bmatrix}
+ \mathcal{O}(\delta^2), \ \delta \to 0.
$$

It follows that approximations for the weights $\mu_1, \ldots, \mu_m$ can be obtained by solving a Vandermonde system.

What happens if we apply the algorithm that we have presented in Chapter 1 in case the zeros of f can be grouped into clusters? The second part of

Theorem 2.1.2 implies that our algorithm stops at $r = m$ if δ, the maximal size of the clusters, is sufficiently small. It returns the zeros of the FOP $\varphi_m(z)$ that is associated with $\langle \cdot, \cdot \rangle$. Theorem 2.1.1 and the fact that the mth degree FOP with respect to $\langle \cdot, \cdot \rangle_m$ is given by $\prod_{j=1}^{m}(z - c_j)$ imply that we can use these zeros as approximations for the centres of the clusters. (This also follows from the first part of Theorem 2.1.2, of course.) The computed approximations for the weights of the clusters should be close to integers. We can check this to verify that we have indeed determined the correct value of m. We can also calculate (approximations for) the ordinary moments associated with $\langle \cdot, \cdot \rangle_m$ and verify if (2.1) is satisfied.

2.2 A numerical example

In the following example we have again considered the case that γ is a circle, cf. Section 1.4 of Chapter 1. The computations have been done via Matlab 5.

Stewart's perturbation theory for the generalized eigenvalue problem [115] allows us to make a sensitivity analysis. An important result from his first order perturbation theory for simple eigenvalues tells us the following. If λ is a simple eigenvalue of the pencil $G_t^{(1)} - \lambda G_t$ and λ_ϵ is the corresponding eigenvalue of a perturbed pencil $\tilde{G}_t^{(1)} - \lambda \tilde{G}_t$ with $\|G_t^{(1)} - \tilde{G}_t^{(1)}\|_2 \approx \|G_t - \tilde{G}_t\|_2 \approx \epsilon$, then

$$\frac{|\lambda - \lambda_\epsilon|}{\sqrt{1 + |\lambda|^2}\sqrt{1 + |\lambda_\epsilon|^2}} \leq \frac{\epsilon}{\sqrt{|y^H G_t^{(1)} x|^2 + |y^H G_t x|^2}} + \mathcal{O}(\epsilon^2)$$
$$=: \kappa(\lambda, x, y)\epsilon + \mathcal{O}(\epsilon^2)$$

where x and y are the right and left eigenvectors corresponding to λ,

$$G_t^{(1)} x = \lambda G_t \quad \text{and} \quad y^H G_t^{(1)} = \lambda y^H G_t,$$

normalized such that $\|x\|_2 = \|y\|_2 = 1$. Let us call $\kappa(\lambda, x, y)$ the *sensitivity factor* of the eigenvalue λ.

Example 2.2.1. Suppose that $n = 10$,

$$z_1 = -1,$$
$$z_2 = 4, \quad z_3 = 4 + \delta(1 + i),$$
$$z_4 = 3i, \quad z_5 = 3i + \delta(10 + 5i), \quad z_6 = 3i + \delta(-3 + 4i),$$
$$z_7 = c + \delta(-1 + 2i), \quad z_8 = c + \delta(1 + 5i), \quad z_9 = c + \delta(1 + i), \quad z_{10} = c + \delta(-2 - 2i),$$

where $c = -3 + 3i$ and $\delta = 10^{-4}$. Suppose that $\nu_1 = \cdots = \nu_{10} = 1$. Let $f(z)$ be the polynomial that has $z_1, \ldots, z_{10}$ as simple zeros, $f(z) = \prod_{k=1}^{10}(z - z_k)$, and let $\gamma = \{z \in \mathbb{C} : |z| = 5\}$. Note that f has four clusters of zeros,

$m = 4$, of weight 1, 2, 3 and 4, respectively. We have evaluated the logarithmic derivative of $f(z)$ again via formula (1.25). As we do not want the algorithm to stop as soon as it has found the approximations for the centres of the clusters, we set ϵ_{stop} to a rather small value, $\epsilon_{stop} = 10^{-18}$. Our algorithm gives the following results. The total number of zeros is equal to 10. The polynomials $\varphi_0(z)$ and $\varphi_1(z)$ are defined as regular FOPs. The computed eigenvalues of the pencil $G_2^{(1)} - \lambda G_2$ lead to zeros that lie inside γ, and thus the polynomial $\varphi_2(z)$ is defined as a regular FOP. The sensitivity factors of the eigenvalues are equal to

```
3.773765047881042e-02
9.474367914383353e-03
```

The solution of the Vandermonde system that corresponds to the approximations for the zeros of $\varphi_2(z)$ is given by

```
7.050510110011339e+00 - i 2.941868186347331e-01
2.949489889988664e+00 + i 2.941868186347336e-01
```

The algorithm computes $\langle \varphi_2(z), \varphi_2(z) \rangle$. In step [4] it compares

```
9.853283227226082e-01
```

with ϵ_{stop} and sets allsmall ← **false**. The polynomial $\varphi_3(z)$ is defined as a regular FOP. The sensitivity factors of the eigenvalues are equal to

```
5.221220847969128e-03
5.691118489264763e-02
9.440130958262826e-03
```

The solution of the Vandermonde system that corresponds to the approximations for the zeros of $\varphi_3(z)$ is given by

```
1.751672615711886e+00 + i 2.391215777821213e-02
5.761474298212248e+00 + i 1.418723240124189e-01
2.486853086075870e+00 - i 1.657844817906288e-01
```

The algorithm computes $\langle \varphi_3(z), \varphi_3(z) \rangle$. In step [4] it compares

```
9.560608076054004e-02
```

with ϵ_{stop} and sets allsmall ← **false**. The polynomial $\varphi_4(z)$ is defined as a regular FOP. The sensitivity factors of the eigenvalues are equal to

```
5.997609084856927e-03
1.632514220886046e-02
2.198497992029341e-02
4.923118129602375e-03
```

The solution of the Vandermonde system that corresponds to the approximations for the zeros of $\varphi_4(z)$ is given by

```
1.999999998746185 e+00 - i  1.260274035855700e-08
4.000000263704519 e+00 - i  1.428161173014011e-07
2.999999744825852 e+00 + i  2.437473983804684e-07
9.999999927234455 e-01 - i  8.832854039794525e-08
```

Observe that these "multiplicities" (actually, they are the weights of the clusters) are at a distance of $\mathcal{O}(10^{-8}) = \mathcal{O}(\delta^2)$ to integers. This is a first indication of the fact that $m = 4$. The algorithm computes $\langle \varphi_4(z), \varphi_4(z) \rangle$. In step [4] it compares

```
4.690246227384357 e-09
```

with ϵ_{stop}. As we have given ϵ_{stop} a very small value, $\epsilon_{stop} = 10^{-18}$, the algorithm sets allsmall $\leftarrow$ **false** and continues. It defines the polynomial $\varphi_5(z)$ as a regular FOP. The sensitivity factors of the eigenvalues are equal to

```
1.859124283121160 e+02
5.997664519337178 e-03
1.632900762729697 e-02
4.930300860498997 e-03
2.201051194773096 e-02
```

Observe that one of the eigenvalues is much more sensitive than the others. The solution of the Vandermonde system that corresponds to the computed approximations for the zeros of $\varphi_5(z)$ is given by

```
-4.028597029147776 e-08 - i  1.692141150994100e-07
 2.000000007231077 e+00 - i  8.308500282558695e-09
 4.000000510541128 e+00 - i  2.651932814361151e-07
 1.000000192451618 e+00 - i  8.614030726982243e-08
 2.999999330062146 e+00 + i  5.288562050783197e-07
```

Observe that the component that corresponds to the spurious eigenvalue is of size $\mathcal{O}(10^{-8})$ whereas the other components are close to integers. This enables us to deduce the presence of spurious eigenvalues without computing the sensitivity factors. The algorithm computes $\langle \varphi_5(z), \varphi_5(z) \rangle$. In step [4] it compares

```
3.544154048709335 e-10
```

with ϵ_{stop} and sets allsmall $\leftarrow$ **false**. The polynomial $\varphi_6(z)$ is defined as a regular FOP. The sensitivity factors of the eigenvalues are equal to

```
1.617034259272598 e+02
1.220513781306331 e+01
5.997660002011287 e-03
1.633002045462034 e-02
4.929209225445833 e-03
2.200115675959345 e-02
```

The solution of the Vandermonde system that corresponds to the approximations for the zeros of $\varphi_6(z)$ is given by

```
-1.109848867030467e-07 - i  3.954458493988473e-08
 5.068148299128812e-12 - i  6.003619075497872e-13
 2.000000007044562e+00 - i  9.500131431430428e-09
 4.000000455291260e+00 - i  3.668720544062149e-07
 1.000000170753607e+00 - i  1.185972639722307e-07
 2.999999477890392e+00 + i  5.345146309560269e-07
```

The algorithm computes $\langle \varphi_6(z), \varphi_6(z) \rangle$. In step [4] it compares

```
1.189012222825083e-09
```

with ϵ_{stop} and sets allsmall $\leftarrow$ **false**. And so on. The algorithm defines $\varphi_7(z)$ as a regular FOP, $\varphi_8(z)$ as an inner polynomial, $\varphi_9(z)$ as a regular FOP, and finally $\varphi_{10}(z)$ as an inner polynomial. The computed approximations for the zeros of f are given by

```
-9.999999999599027e-01 + i  1.827471507453993e-11
 4.000066008247924e+00 + i  5.012986226260452e-05
 4.000047628710319e+00 - i  2.622572865154105e-04
-5.868580001572310e-05 + i  3.000189455314092e+00
 1.147726665656712e-03 + i  3.000342939244789e+00
-3.000001882960546e+00 + i  3.000324337949845e+00
-3.000258740452308e+00 + i  2.999857462002724e+00
-9.085700394437596e-01 + i  1.866967988946470e+00
-8.509233181288104e-01 - i  6.684482268880147e+00
-4.999300000000002e-01 + i  2.100160000000000e+00
```

The relative errors of the approximations for the zeros that belong to the clusters of weight 1 and 2 are $\mathcal{O}(10^{-11})$ and $\mathcal{O}(10^{-5})$, resp. For the other zeros, the relative errors are at least $\mathcal{O}(10^{-3})$.

If we set $\epsilon_{stop} = 10^{-6}$, then our algorithm stops at the polynomial of degree 4. We obtain the following approximations for the centres of the clusters:

```
-9.999999564181510e-01 - i  5.152524762408461e-08
 4.000050001653271e+00 + i  5.000694739720757e-05
 2.335838430156945e-04 + i  3.000299920075392e+00
-3.000024926663507e+00 + i  3.000149946356108e+00
```

Let us now focus on the separate clusters. We have considered the circles whose centre is the computed approximation for the centre of a cluster and whose radius is equal to 0.1. The relative errors of the approximations for the zeros that we obtain are $\mathcal{O}(10^{-16})$, $\mathcal{O}(10^{-12})$, $\mathcal{O}(10^{-10})$ and $\mathcal{O}(10^{-6})$ for the clusters of weight 1, 2, 3 and 4, respectively. If we consider a circle whose centre is the computed approximation for the centre of the cluster of weight 4 and whose radius is equal to 10^{-3}, then the relative errors of the approximations that we obtain for the zeros that lie in this cluster are $\mathcal{O}(10^{-16})$. $\diamond$

More numerical examples will be given in Section 2.4.

2.3 Rational interpolation at roots of unity

We will now approach our problem of computing all the zeros of f that lie inside γ in a different way, based on rational interpolation at roots of unity. We will show how the new approach complements the previous one and how it can be used effectively in case γ is the unit circle. Numerical examples will be given in Section 2.4.

Let K be a positive integer and let $t_1, \ldots, t_K$ be the Kth roots of unity,

$$t_k := \exp\left(\frac{2\pi i}{K}k\right), \qquad k = 1, \ldots, K.$$

Define $g_{K-1}(z)$ as the polynomial

$$g_{K-1}(z) := s_0 z^{K-1} + s_1 z^{K-2} + \cdots + s_{K-1}.$$

Note that $\deg g_{K-1}(z) = K - 1$ as by assumption $s_0 \neq 0$. Without loss of generality we may assume that $g_{K-1}(t_k) \neq 0$ for $k = 1, \ldots, K$. (This condition will be needed in Theorem 2.3.1.)

Let $w_K(z) := z^K - 1$ and define the symmetric bilinear form $\langle\langle \cdot, \cdot \rangle\rangle$ as

$$\langle\langle \phi, \psi \rangle\rangle := \sum_{k=1}^{K} \frac{g_{K-1}(t_k)}{w_K'(t_k)} \phi(t_k)\psi(t_k)$$

for $\phi, \psi \in \mathcal{P}$. Note that this form can be evaluated via FFT.

Define the ordinary moments σ_p associated with the form $\langle\langle \cdot, \cdot \rangle\rangle$ as

$$\sigma_p := \langle\langle 1, z^p \rangle\rangle = \sum_{k=1}^{K} \frac{g_{K-1}(t_k)}{w_K'(t_k)} t_k^p$$

for $p = 0, 1, 2, \ldots$ and let $\mathcal{H}_k$ be the $k \times k$ Hankel matrix

$$\mathcal{H}_k := \left[\sigma_{p+q}\right]_{p,q=0}^{k-1} = \begin{bmatrix} \sigma_0 & \sigma_1 & \cdots & \sigma_{k-1} \\ \sigma_1 & & \cdot^{\cdot^{\cdot}} & \vdots \\ \vdots & \cdot^{\cdot^{\cdot}} & & \vdots \\ \sigma_{k-1} & \cdots & \cdots & \sigma_{2k-2} \end{bmatrix}$$

for $k = 1, 2, \ldots$. Then the regular FOP f_τ of degree $\tau \geq 1$ associated with the form $\langle\langle \cdot, \cdot \rangle\rangle$ exists if and only if the matrix $\mathcal{H}_\tau$ is nonsingular. Also, the following theorem holds (cf. Theorem 1.2.1).

Theorem 2.3.1. $K = \operatorname{rank} \mathcal{H}_{K+p}$ *for every nonnegative integer* p.

Thus $\mathcal{H}_K$ is nonsingular whereas $\mathcal{H}_\tau$ is singular for $\tau > K$. The regular FOP f_K of degree K exists while regular FOPs of degree larger than K do not exist. The polynomial f_K is easily seen to be

$$f_K(z) = (z - t_1) \cdot \cdots \cdot (z - t_K) = w_K(z).$$

It is the monic polynomial of degree K that has $t_1, \ldots, t_K$ as simple zeros.

If $\mathcal{H}_K$ is strongly nonsingular, then we have a full set $\{f_0, f_1, \ldots, f_K\}$ of regular FOPs. Else, we can proceed in the same way as with the form $\langle \cdot, \cdot \rangle$. By filling up the gaps in the sequence of existing regular FOPs it is possible to define a sequence $\{f_\tau\}_{\tau=0}^\infty$, with f_τ a monic polynomial of degree τ, such that if these polynomials are grouped into blocks according to the sequence of regular indices, then polynomials belonging to different blocks are orthogonal with respect to $\langle\langle \cdot, \cdot \rangle\rangle$. More precisely, define $\{f_\tau\}_{\tau=0}^\infty$ as follows. If τ is a regular index, then let f_τ be the regular FOP of degree τ. Else define f_τ as $f_\rho p_{\tau,\rho}$ where ρ is the largest regular index less than τ and $p_{\tau,\rho}$ is an arbitrary monic polynomial of degree $\tau - \rho$. In the latter case f_τ is called an *inner polynomial*. If $p_{\tau,\rho}(z) = z^{\tau-\rho}$ then we say that f_τ is defined *by using the standard monomial basis*. The block orthogonality property is expressed by the fact that the Gram matrix $[\langle\langle f_r, f_s \rangle\rangle]_{r,s=0}^{K-1}$ is block diagonal. The diagonal blocks are nonsingular, symmetric and zero above the main antidiagonal. If all the inner polynomials in a certain block are defined by using the standard monomial basis, then the corresponding diagonal block has Hankel structure, cf. Theorem 1.2.3.

The definition of the form $\langle\langle \cdot, \cdot \rangle\rangle$ may seem arbitrary. However, there exists a remarkable connection between the forms $\langle \cdot, \cdot \rangle$ and $\langle\langle \cdot, \cdot \rangle\rangle$.

Theorem 2.3.2. Let $\phi, \psi \in \mathcal{P}$. If $\deg \phi + \deg \psi \leq K - 1$, then $\langle\langle \phi, \psi \rangle\rangle = \langle \phi, \psi \rangle$.

Proof. Let $V(t_1, \ldots, t_K)$ be the Vandermonde matrix with nodes $t_1, \ldots, t_K$,

$$V(t_1, \ldots, t_K) := \begin{bmatrix} 1 & t_1 & \cdots & t_1^{K-1} \\ \vdots & \vdots & & \vdots \\ 1 & t_K & \cdots & t_K^{K-1} \end{bmatrix}.$$

Then

$$\begin{bmatrix} g_{K-1}(t_1) \\ \vdots \\ g_{K-1}(t_K) \end{bmatrix} = V(t_1, \ldots, t_K) \begin{bmatrix} s_{K-1} \\ \vdots \\ s_0 \end{bmatrix}.$$

As $V(t_1, \ldots, t_K)/\sqrt{K}$ is unitary, it follows that

$$\begin{bmatrix} s_{K-1} \\ \vdots \\ s_0 \end{bmatrix} = \frac{1}{K} [V(t_1, \ldots, t_K)]^H \begin{bmatrix} g_{K-1}(t_1) \\ \vdots \\ g_{K-1}(t_K) \end{bmatrix}.$$

As $w_K(z) = z^K - 1$, it follows that $w_K'(z) = Kz^{K-1}$ and thus $w_K'(t_k) = K/t_k$ for $k = 1, \ldots, K$. Let $j \in \{1, \ldots, K\}$. Then

$$\sigma_{K-j} = \frac{1}{K} \sum_{k=1}^{K} g_{K-1}(t_k) t_k^{K-j+1} = \frac{1}{K} \sum_{k=1}^{K} g_{K-1}(t_k) \overline{t_k^{j-1}}$$

and thus

$$\begin{bmatrix} \sigma_{K-1} \\ \sigma_{K-2} \\ \vdots \\ \sigma_0 \end{bmatrix} = \frac{1}{K} \overline{\begin{bmatrix} 1 & \cdots & 1 \\ t_1 & \cdots & t_K \\ \vdots & & \vdots \\ t_1^{K-1} & \cdots & t_K^{K-1} \end{bmatrix}} \begin{bmatrix} g_{K-1}(t_1) \\ g_{K-1}(t_2) \\ \vdots \\ g_{K-1}(t_K) \end{bmatrix}$$

$$= \frac{1}{K} [V(t_1, \ldots, t_K)]^H \begin{bmatrix} g_{K-1}(t_1) \\ \vdots \\ g_{K-1}(t_K) \end{bmatrix}$$

$$= \begin{bmatrix} s_{K-1} \\ s_{K-2} \\ \vdots \\ s_0 \end{bmatrix}.$$

In other words, $s_p = \sigma_p$ for $p = 0, 1, \ldots, K-1$. As $\langle\langle \phi, \psi \rangle\rangle$ depends on σ_p for $p = 0, 1, \ldots, \deg(\phi\psi)$, this proves the theorem.   □

Corollary 2.3.1. *Let τ be a nonnegative integer. If $2\tau - 1 \leq K$, then τ is a regular index for $\langle\langle \cdot, \cdot \rangle\rangle$ if and only if τ is a regular index for $\langle \cdot, \cdot \rangle$. Moreover, if $2\tau \leq K$ and if τ is a regular index, then $f_\tau(z) \equiv \varphi_\tau(z)$. Else, if τ is not a regular index, then $f_\tau(z) = R_{\tau,\rho}(z)\varphi_\tau(z)$ where ρ is the largest regular index less than τ and $R_{\tau,\rho}(z)$ is a rational function of type $[\tau - \rho/\tau - \rho]$. If $f_\tau(z)$ and $\varphi_\tau(z)$ are both defined by using the standard monomial basis, then $R_{\tau,\rho}(z) \equiv 1$.*

Corollary 2.3.2. *If $K \geq 2n$ and $n \leq \tau \leq \lfloor K/2 \rfloor$, then $f_\tau(z_k) = 0$ for $k = 1, \ldots, n$ and $\langle z^p, f_\tau(z) \rangle = 0$ for all $p \geq 0$. Also, $\langle\langle z^p, f_\tau(z) \rangle\rangle = 0$ for $p = \tau, \ldots, K - 1 - \tau$. (Note that the latter range may be empty.)*

Corollary 2.3.3. *If $K \geq 2m$ and $m \leq \tau \leq \lfloor K/2 \rfloor$, then $f_\tau(c_j) = \mathcal{O}(\delta^2)$, $\delta \to 0$ for $j = 1, \ldots, m$ and $\langle z^p, f_\tau(z) \rangle = \mathcal{O}(\delta^2)$, $\delta \to 0$ for all $p \geq \tau$. Also, $\langle\langle z^p, f_\tau(z) \rangle\rangle = \mathcal{O}(\delta^2)$, $\delta \to 0$ for $p = \tau, \ldots, K - 1 - \tau$. (Note that the latter range may be empty.)*

Thus, if $K \geq 2N$, then we can apply the algorithm that we have presented in Chapter 1 to the form $\langle\langle \cdot, \cdot \rangle\rangle$ and we will obtain exactly the same results as with the form $\langle \cdot, \cdot \rangle$. This is an interesting fact in its own right. The main reason, though, that motivated us to introduce the form $\langle\langle\cdot,\cdot\rangle\rangle$ is the fact that it is related to rational interpolation. We will show that the "denominator polynomials" in a certain linearized rational interpolation problem that is related to the polynomial $g_{K-1}(z)$ are FOPs with respect to $\langle\langle\cdot,\cdot\rangle\rangle$. This will lead to an alternative way to calculate the FOPs $f_\tau(z)$ and thus, because of Corollary 2.3.1, the FOPs $\varphi_\tau(z)$.

Let σ and τ be nonnegative integers such that $\sigma + \tau + 1 = K$. Let $p_\sigma(z)$ and $q_\tau(z)$ be polynomials, where

$$\deg p_\sigma(z) \le \sigma \quad \text{and} \quad \deg q_\tau(z) \le \tau, \tag{2.2}$$

such that the following linearized rational interpolation conditions are satisfied:

$$p_\sigma(t_k) - q_\tau(t_k)g_{K-1}(t_k) = 0, \qquad k = 1,\ldots,K. \tag{2.3}$$

Each pair of polynomials $(p_\sigma(z), q_\tau(z))$ that satisfies the degree conditions (2.2) and the interpolation conditions (2.3) is called a *multipoint Padé form* (MPF). The polynomials $p_\sigma(z)$ and $q_\tau(z)$ will be called *numerator polynomial* and *denominator polynomial*, respectively.

The interpolation conditions (2.3) lead to a system of K linear equations in $K + 1$ unknowns, and thus at least one nontrivial (i.e., whose numerator and denominator polynomial are not identically equal to zero) MPF exists. As (2.3) are homogeneous linear equations, every scalar multiple of a MPF is also a MPF. From now on, we will always assume that MPFs are normalized such that the denominator polynomial is monic. However, the fact that then the number of interpolation conditions is equal to the number of unknown polynomial coefficients, does not guarantee that there exists only one MPF. It merely guarantees that every MPF leads to the same irreducible rational function, called *multipoint Padé approximant* (MPA). Indeed, suppose that there exist two MPAs. Then the numerator polynomial of the difference of these MPAs is a polynomial of degree $\le \sigma + \tau$ that vanishes at $\sigma + \tau + 1$ points. This numerator polynomial is therefore identically equal to zero, which implies that the MPA is unique.

Let $\mathcal{R}_{\sigma,\tau}$ be the set of rational functions of type $[\sigma/\tau]$, i.e., with numerator degree at most σ and denominator degree at most τ. A rational interpolation problem that is closely related to (2.3) is the *Cauchy interpolation problem*: find all irreducible rational functions $r_{\sigma,\tau}(z) \in \mathcal{R}_{\sigma,\tau}$ whose denominator polynomial is monic, such that

$$r_{\sigma,\tau}(t_k) = g_{K-1}(t_k), \qquad k = 1,\ldots,K. \tag{2.4}$$

This interpolation problem is not always solvable. If a solution exists, then it is unique, and it is equal to the MPA. In general, however, the MPA need not solve the Cauchy interpolation problem: the numerator and denominator polynomials of the MPFs may have common zeros at some interpolation points. The MPA may not satisfy the interpolation condition (2.4) at these points, which are then called *unattainable points*.

Let $r(z) \in \mathcal{R}_{\sigma,\tau}$ and suppose that $r(z) = p(z)/q(z)$ where $p(z)$ and $q(z)$ are relatively prime polynomials. The *defect* of r with respect to $\mathcal{R}_{\sigma,\tau}$ is then defined as

$$\min\{\, \sigma - \deg p(z), \tau - \deg q(z) \,\}.$$

The following theorem provides the general solution of the linearized rational interpolation problem.

Theorem 2.3.3. *The general MPF that corresponds to the degree conditions (2.2) and the interpolation conditions (2.3) is given by*

$$\big(p_\sigma(z), q_\tau(z)\big) = \big(\hat{p}_\sigma(z)s(z)u(z), \hat{q}_\tau(z)s(z)u(z)\big),$$

where $\hat{p}_\sigma(z)$, $\hat{q}_\tau(z)$ and $s(z)$ are uniquely determined polynomials, and where $u(z)$ is arbitrary. The polynomials $\hat{p}_\sigma(z)$ and $\hat{q}_\tau(z)$ are relatively prime, and $s(z)$ is a divisor of $w_K(z)$. Let $\hat{\delta}_{\sigma,\tau}$ be the defect of $\hat{p}_\sigma(z)/\hat{q}_\tau(z)$ with respect to $\mathcal{R}_{\sigma,\tau}$. Then $\deg s(z) \leq \hat{\delta}_{\sigma,\tau}$ and $\deg u(z) \leq \hat{\delta}_{\sigma,\tau} - \deg s(z)$. The zeros of $s(z)$ are the unattainable points for the corresponding Cauchy interpolation problem.

Proof. See, for example, Gutknecht [62, p. 549]. $\square$

The literature on rational interpolation is vast. We will not give a full account of all the other issues (in particular, the block structure of the Newton-Padé table) that are involved. The reader may wish to consult the papers by Meinguet [99], Antoulas [9, 10, 11], Berrut and Mittelmann [16] or Gutknecht [62, 63, 64, 65], and the references cited therein.

What is of special interest to us, is the fact that the denominator polynomials $q_\tau(z)$ are formal orthogonal polynomials with respect to $\langle\langle \cdot, \cdot \rangle\rangle$.

Theorem 2.3.4. *Let σ and τ be nonnegative integers such that $\sigma + \tau + 1 = K$. Let $(p_\sigma(z), q_\tau(z))$ be a MPF for the degree conditions (2.2) and the interpolation conditions (2.3). Then $\langle\langle z^p, q_\tau(z) \rangle\rangle = 0$ for $p = 0, 1, \dots, K - 2 - \deg p_\sigma(z)$ and $\langle\langle z^p, q_\tau(z) \rangle\rangle \neq 0$ if $p = K - 1 - \deg p_\sigma(z)$.*

Proof. Apparently this orthogonality relation was already known to Jacobi. As it plays a very important role in our approach, we prefer to give a (short, but explicit) proof. See also Eğecioğlu and Koç [39] and Gemignani [57] for a slightly weaker version of this theorem.

Let $p \in \{0, 1, \dots, K - 2 - \deg p_\sigma(z)\}$. Then

$$\sum_{k=1}^{K} \frac{t_k^p p_\sigma(t_k)}{w_K'(t_k)} = \sum_{k=1}^{K} \frac{g_{K-1}(t_k)}{w_K'(t_k)} t_k^p q_\tau(t_k) \tag{2.5}$$

and $z^p p_\sigma(z)$ is a polynomial of degree $p + \deg p_\sigma(z) \leq K - 2$. Lagrange's formula for the polynomial $y_{K-1}(z)$ of degree $\leq K - 1$ that interpolates the polynomial $z^p p_\sigma(z)$ in the points $t_1, \ldots, t_K$ implies that the left hand side of (2.5) is equal to the coefficient of z^{K-1} of $y_{K-1}(z)$. As $\deg[z^p p_\sigma(z)] < K-1$, it follows that $y_{K-1}(z) \equiv z^p p_\sigma(z)$ and that the coefficient of z^{K-1} of $y_{K-1}(z)$ is equal to zero. It follows that $\langle\langle z^p, q_\tau(z) \rangle\rangle = 0$ for $p = 0, 1, \ldots, K - 2 - \deg p_\sigma(z)$. A similar reasoning shows that $\langle\langle z^p, q_\tau(z) \rangle\rangle \neq 0$ if $p = K - 1 - \deg p_\sigma(z)$. This proves the theorem. $\square$

The following theorem implies that the coefficients (in the standard monomial basis) of the numerator polynomial $p_\sigma(z)$ of a MPF $\big(p_\sigma(z), q_\tau(z)\big)$ that corresponds to the degree conditions (2.2) and the interpolation conditions (2.3) can be expressed as inner products with respect to $\langle\langle \cdot, \cdot \rangle\rangle$. This explains how the degree property $\deg p_\sigma(z) \leq \sigma$ is related to the formal orthogonality property satisfied by $q_\tau(z)$.

Theorem 2.3.5. *Suppose that $q(z)$ is a polynomial and let $p(z)$ be the polynomial of degree $\leq K - 1$ that interpolates $g_{K-1}(z)q(z)$ at the points $t_1, \ldots, t_K$. Let $p(z) =: p_0 + p_1 z + \cdots + p_{K-1} z^{K-1}$. Then $p_k = \langle\langle z^{K-1-k}, q(z) \rangle\rangle$ for $k = 0, 1, \ldots, K - 1$.*

Proof. The Lagrange representation of $p(z)$ is given by

$$p(z) = \sum_{k=1}^{K} \pi_k L_k(z)$$

where

$$\pi_k := \frac{g_{K-1}(t_k)q(t_k)}{w'_K(t_k)} \quad \text{and} \quad L_k(z) := \frac{w_K(z)}{z - t_k}$$

for $k = 1, \ldots, K$. Let $L_k(z) =: L_{0,k} + L_{1,k} z + \cdots + L_{K-1,k} z^{K-1}$ for $k = 1, \ldots, K$. Note that $L_{K-1,1} = \cdots = L_{K-1,K} = 1$. Let

$$V := \begin{bmatrix} 1 & t_1 & \cdots & t_1^{K-1} \\ \vdots & \vdots & & \vdots \\ 1 & t_K & \cdots & t_K^{K-1} \end{bmatrix}$$

be the Vandermonde matrix with nodes $t_1, \ldots, t_K$ and let

$$L := \begin{bmatrix} L_{0,1} & \cdots & L_{0,K} \\ \vdots & & \vdots \\ L_{K-1,1} & \cdots & L_{K-1,K} \end{bmatrix}$$

be the matrix that contains the coefficients of $L_1(z), \ldots, L_K(z)$. Then

$$VL = \text{diag}\,(L_1(t_1), \ldots, L_K(t_K))$$
$$= \text{diag}\,(w'_K(t_1), \ldots, w'_K(t_K))$$
$$= K\,\text{diag}\,(\overline{t_1}, \ldots, \overline{t_K}).$$

As $V/\sqrt{K}$ is unitary, it follows that $V^{-1} = V^H/K$ and thus

$$L = V^H \text{diag}\,(\overline{t_1}, \ldots, \overline{t_K}) = \begin{bmatrix} t_1^{K-1} & \cdots & t_K^{K-1} \\ \vdots & & \vdots \\ t_1 & \cdots & t_K \\ 1 & \cdots & 1 \end{bmatrix}.$$

In other words, $L_{j,k} = t_k^{K-1-j}$ for $k = 1, \ldots, K$ and $j = 0, 1, \ldots, K - 1$. As $p_j = \sum_{k=1}^{K} L_{j,k} \pi_k$ for $j = 0, 1, \ldots, K - 1$, it follows that

$$p_j = \sum_{k=1}^{K} \frac{g_{K-1}(t_k)}{w'_K(t_k)} t_k^{K-1-j} q(t_k) = \langle\langle z^{K-1-j}, q(z) \rangle\rangle$$

for $j = 0, 1, \ldots, K - 1$. This proves the theorem. $\qquad\qquad\square$

The following theorem shows how to construct the sequence of FOPs $f_\tau(z)$ from the MPFs $(p_\sigma(z), q_\tau(z))$. Regular FOPs correspond to denominator polynomials whose degree is equal to τ whereas the sizes of the blocks are determined by the actual degrees of the numerator polynomials.

Theorem 2.3.6. *Let σ and τ be nonnegative integers such that $\sigma + \tau + 1 = K$. Let $(p_\sigma(z), q_\tau(z)) = (\hat{p}_\sigma(z)s(z)u(z), \hat{q}_\tau(z)s(z)u(z))$ be the general MPF for the degree conditions (2.2) and the interpolation conditions (2.3), where the polynomials $\hat{p}_\sigma(z)$, $\hat{q}_\tau(z)$, $s(z)$ and $u(z)$ are as in Theorem 2.3.3. If τ is a regular index for $\langle\langle \cdot, \cdot \rangle\rangle$, then $\deg(\hat{q}_\tau(z)s(z)) = \tau$, the FOP $f_\tau(z)$ is given by $f_\tau(z) \equiv \hat{q}_\tau(z)s(z)$ and the smallest regular index that is larger than τ is equal to $K - \deg(\hat{p}_\sigma(z)s(z))$. Conversely, if $\deg(\hat{q}_\tau(z)s(z)) = \tau$, then τ is a regular index for $\langle\langle \cdot, \cdot \rangle\rangle$.*

Proof. Suppose that τ is a regular index for $\langle\langle \cdot, \cdot \rangle\rangle$. Then $\det \mathcal{H}_\tau \neq 0$ and there exists precisely one monic polynomial $f_\tau(z)$ of degree τ such that

$$\langle\langle z^p, f_\tau(z) \rangle\rangle = 0 \qquad \text{for } p = 0, 1, \ldots, \tau - 1.$$

Let $p(z)$ be the polynomial of degree $\leq K - 1$ that interpolates $g_{K-1}(z)f_\tau(z)$ at the points $t_1, \ldots, t_K$. Then, according to Theorem 2.3.5, $\deg p(z) \leq K - \tau - 1 = \sigma$. Thus $(p(z), f_\tau(z))$ is a MPF for the degree conditions (2.2) and the interpolation conditions (2.3). In other words, there exists a MPF whose denominator polynomial has degree τ. Theorem 2.3.3 then implies that there exists a monic polynomial $u_\tau(z)$ of degree $\tau - \deg(\hat{q}_\tau(z)s(z))$ such that $f_\tau(z) \equiv \hat{q}_\tau(z)s(z)u_\tau(z)$. If $\deg u_\tau(z) > 0$, then we can choose a different monic polynomial $\tilde{u}_\tau(z)$ of the same degree. Then $f_\tau(z) \not\equiv \hat{q}_\tau(z)s(z)\tilde{u}_\tau(z)$ and, by Theorem 2.3.4,

$$\langle\langle z^p, \hat{q}_\tau(z)s(z)\tilde{u}_\tau(z)\rangle\rangle = 0 \qquad \text{for } p = 0, 1, \ldots, \tau - 1.$$

As $\deg(\hat{q}_\tau(z)s(z)\tilde{u}_\tau(z)) = \tau$, this contradicts the fact that $f_\tau(z)$ is unique. Thus we may conclude that $\deg(\hat{q}_\tau(z)s(z)) = \tau$ and $f_\tau(z) \equiv \hat{q}_\tau(z)s(z)$. Now Theorem 2.3.4 implies that

$$\langle\langle z^p, f_\tau(z)\rangle\rangle = 0 \qquad \text{for } p = 0, 1, \ldots, K - 2 - \deg(\hat{p}_\sigma(z)s(z))$$

and

$$\langle\langle z^p, f_\tau(z)\rangle\rangle \neq 0 \qquad \text{if } p = K - 1 - \deg(\hat{p}_\sigma(z)s(z)).$$

The structure of the diagonal blocks of the Gram matrix $[\langle\langle f_r, f_s\rangle\rangle]_{r,s=0}^{K-1}$ then implies that $\det \mathcal{H}_t = 0$ for $t = \tau + 1, \ldots, K - 1 - \deg(\hat{p}_\sigma(z)s(z))$ and that $\det \mathcal{H}_t \neq 0$ if $t = K - \deg(\hat{p}_\sigma(z)s(z))$.

Suppose that $\deg(\hat{q}_\tau(z)s(z)) = \tau$. Then there exists only one MPF for the degree conditions (2.2) and the interpolation conditions (2.3). The polynomial $\hat{q}_\tau(z)s(z)$ is a monic polynomial of degree τ and, according to Theorem 2.3.4,

$$\langle\langle z^p, \hat{q}_\tau(z)s(z)\rangle\rangle = 0 \qquad \text{for } p = 0, 1, \ldots, \tau - 1.$$

Suppose that there exists another monic polynomial $\tilde{f}_\tau(z)$ of degree τ, $\tilde{f}_\tau(z) \not\equiv \hat{q}_\tau(z)s(z)$, such that

$$\langle\langle z^p, \tilde{f}_\tau(z)\rangle\rangle = 0 \qquad \text{for } p = 0, 1, \ldots, \tau - 1.$$

Let $p(z)$ be the polynomial of degree $\leq K - 1$ that interpolates $g_{K-1}(z)\tilde{f}_\tau(z)$ at the points $t_1, \ldots, t_K$. Then, according to Theorem 2.3.5, $\deg p(z) \leq K - \tau - 1 = \sigma$. Thus $(p(z), \tilde{f}_\tau(z))$ is a MPF for the degree conditions (2.2) and the interpolation conditions (2.3). It follows that $\tilde{f}_\tau(z) \equiv \hat{q}_\tau(z)s(z)$. In other words, there exists only one monic polynomial of degree τ that is orthogonal (with respect to $\langle\langle \cdot, \cdot \rangle\rangle$) to all polynomials of lower degree. Thus τ is a regular index for $\langle\langle \cdot, \cdot \rangle\rangle$. This proves the theorem.   □

The previous theorem suggests the following look-ahead strategy. Start with $\tau = 0$ and the corresponding MPF $(g_{K-1}(z), 1)$. Then set $\tau \leftarrow K - \deg g_{K-1}(z)$. Note that $\tau = 1$ as $\deg g_{K-1}(z) = K - 1$. Compute the corresponding MPF $(p_\sigma(z), q_\tau(z))$. Note that, as τ is a regular index for $\langle\langle \cdot, \cdot \rangle\rangle$, this MPF is uniquely defined, i.e., the polynomial $u(z) \equiv 1$ (cf. Theorem 2.3.3). Use $K - \deg p_\sigma(z)$ as the next value of τ, and so on. Observe that, if $\deg p_\sigma(z) = \sigma$, then the next value of τ is given by $\tau + 1$. The interpolation problems can be solved via the algorithm of Van Barel and Bultheel [119]. This algorithm provides the coefficients of the numerator and the denominator polynomial in the standard monomial basis.

Of course, in floating-point arithmetic this strategy will only work if one uses a concept of 'numerical degree' instead of the classical 'degree'. The numerical degree of a polynomial can be defined as follows. Let $\epsilon > 0$. The ϵ-degree of a polynomial $p(z) =: p_0 + p_1 z + \cdots + p_{K-1} z^{K-1} \in \mathcal{P}$ of degree $\leq K - 1$ is defined as follows. Let

$$\chi(k) := \frac{\max\{|p_{k+1}|, \dots, |p_{K-1}|\}}{|p_k|}$$

for all $k \in \{0, 1, \dots, K - 2\}$ such that $p_k \neq 0$ and $\chi(k) := \infty$ otherwise. If

$$\min_{0 \leq k \leq K-2} \chi(k) \leq \epsilon, \tag{2.6}$$

then the ϵ-degree of $p(z)$ is defined as the index k for which the minimum in (2.6) is attained. Else, the ϵ-degree of $p(z)$ is set equal to $K - 1$.

The following corollaries provide us with a stopping criterion.

Corollary 2.3.4. *Let σ and τ be nonnegative integers such that $\sigma + \tau + 1 = K$. Let $(p_\sigma(z), q_\tau(z))$ be a MPF for the degree conditions (2.2) and the interpolation conditions (2.3). If $K \geq 2n$ and $n \leq \tau \leq \lfloor K/2 \rfloor$, then $\deg p_\sigma(z) \leq \deg q_\tau(z) - 1$.*

Proof. This follows from Theorem 2.3.5, Theorem 2.3.2, and Corollary 2.3.2. □

Corollary 2.3.5. *Let σ and τ be nonnegative integers such that $\sigma + \tau + 1 = K$. Let $(p_\sigma(z), q_\tau(z))$ be a MPF for the degree conditions (2.2) and the interpolation conditions (2.3). Let $p_\sigma(z) =: p_0 + p_1 z + \cdots + p_{K-1} z^{K-1}$. If $K \geq 2n$ and $m \leq \tau \leq \lfloor K/2 \rfloor$, then $p_k = \mathcal{O}(\delta^2)$, $\delta \to 0$ for $k = \deg q_\tau(z), \dots, K - 1$. In other words, if ϵ is sufficiently small, then the ϵ-degree of $p_\sigma(z)$ is less or equal than $\deg q_\tau(z) - 1$.*

Proof. This follows from Theorem 2.3.5, Theorem 2.3.2, and Corollary 2.3.3. □

In other words, at the end the (numerical) degree of the numerator polynomial is less or equal than the degree of the denominator polynomial minus one. One can easily verify that this stopping criterion is equivalent to the one used in the algorithm that we have presented in Chapter 1.

Let us consider the problem of how to evaluate the polynomial $g_{K-1}(z)$ at the Kth roots of unity $t_1, \dots, t_K$. One can easily verify that

$$g_{K-1}(z) = \frac{1}{2\pi i} \int_\gamma \frac{t^K - z^K}{t - z} \frac{f'(t)}{f(t)} \, dt$$

if $z \notin \gamma$. Thus, if $t_k \notin \gamma$ for $k = 1, \dots, K$, then

$$g_{K-1}(t_k) = \frac{1}{2\pi i} \int_\gamma \frac{t^K - 1}{t - t_k} \frac{f'(t)}{f(t)} dt$$

for $k = 1, \ldots, K$.

In case γ is the unit circle, one can obtain accurate approximations for

$$g_{K-1}(t_1), \ldots, g_{K-1}(t_K)$$

in a very efficient way. Let L be a positive integer $\geq K$, preferably a power of 2. Let $\omega_1, \ldots, \omega_L$ be the Lth roots of unity,

$$\omega_l := \exp\left(\frac{2\pi i}{L} l\right), \qquad l = 1, \ldots, L.$$

Theorem 2.3.7. *Suppose that γ is the unit circle. Let $v_L \in \mathbb{C}^{L \times 1}$ be the vector*

$$v_L := \frac{1}{L} \begin{bmatrix} 1 & \omega_1 & \cdots & \omega_1^{L-1} \\ \vdots & \vdots & & \vdots \\ 1 & \omega_L & \cdots & \omega_L^{L-1} \end{bmatrix}^H \begin{bmatrix} (f'/f)(\omega_1) \\ \vdots \\ (f'/f)(\omega_L) \end{bmatrix}.$$

Then

$$\begin{bmatrix} O_{K \times (L-K)} & I_K \end{bmatrix} v_L \approx \begin{bmatrix} s_{K-1} \\ \vdots \\ s_0 \end{bmatrix}$$

where $O_{K \times (L-K)}$ denotes the $K \times (L - K)$ zero matrix and I_K denotes the $K \times K$ identity matrix. In other words, the K last components of v_L are approximations for $s_{K-1}, \ldots, s_1, s_0$.

Proof. By approximating

$$s_p = \frac{1}{2\pi i} \int_\gamma z^p \frac{f'(z)}{f(z)} dz = \frac{1}{2\pi} \int_0^{2\pi} e^{ip\theta} e^{i\theta} \frac{f'(e^{i\theta})}{f(e^{i\theta})} d\theta, \qquad p = 0, 1, 2, \ldots,$$

via the trapezoidal rule, we obtain that

$$s_p \approx \frac{1}{L} \sum_{l=1}^{L} \omega_l^{p+1} \frac{f'(\omega_l)}{f(\omega_l)}, \qquad p = 0, 1, 2, \ldots.$$

It follows that

$$s_p \approx \frac{1}{L} \sum_{l=1}^{L} \overline{\omega_l}^{L-1-p} \frac{f'(\omega_l)}{f(\omega_l)}, \qquad p = 0, 1, \ldots, L - 1.$$

This proves the theorem. $\qquad\qquad\qquad\qquad\qquad\qquad\qquad\qquad\qquad\square$

Since

$$
\begin{bmatrix} g_{K-1}(t_1) \\ \vdots \\ g_{K-1}(t_K) \end{bmatrix} = \begin{bmatrix} 1 & t_1 & \cdots & t_1^{K-1} \\ \vdots & \vdots & & \vdots \\ 1 & t_K & \cdots & t_K^{K-1} \end{bmatrix} \begin{bmatrix} s_{K-1} \\ \vdots \\ s_0 \end{bmatrix},
$$

it follows that we can obtain approximations for $g_{K-1}(t_1), \ldots, g_{K-1}(t_K)$ via one L-point (inverse) FFT and one K-point FFT.

2.4 More numerical examples

We have implemented the strategy described in the previous section in Matlab. We have considered the case that γ is the unit circle. Approximations for

$$
g_{K-1}(t_1), \ldots, g_{K-1}(t_K)
$$

have been computed by using Theorem 2.3.7. The interpolation problems have been solved via the algorithm of Van Barel and Bultheel [119].

Example 2.4.1. Let us reconsider the problem that we have studied in Example 2.2.1. As the corresponding γ is given by $\gamma = \{ z \in \mathbb{C} : |z| = 5 \}$, we divide all the zeros by 5 to transform the problem to the unit disk. Recall that $N = n = 10$ whereas $m = 4$. We set $L = 512$ and $K = 22$.

In Figure 2.1 we plot the logarithm with base 10 of the modulus of the coefficients of $p_\sigma(z)$ for $\tau = 0, 1, \ldots, 5$. (The logarithm of the modulus of the lowest degree coefficient is shown on the left. In general, the coefficient of z^k corresponds to the abscis $k + 1$.) Note that $\sigma = 21 - \tau$. Clearly $m = 4$. By multiplying the zeros of $q_4(z)$, as computed via the Matlab command roots, by 5 to transform them back to the setting of Example 2.2.1, we obtain the following:

```
-9.999999564178451e-01 - i  5.152536484956870e-08
 4.000050001653282e+00 + i  5.000694740214643e-05
 2.335838430343232e-04 + i  3.000299920075351e+00
-3.000024926663527e+00 + i  3.000149946356137e+00
```

These values are to be compared with the approximations for the centres of the clusters that we have obtained in Example 2.2.1, namely

```
-9.999999564181510e-01 - i  5.152524762408461e-08
 4.000050001653271e+00 + i  5.000694739720757e-05
 2.335838430156945e-04 + i  3.000299920075392e+00
-3.000024926663507e+00 + i  3.000149946356108e+00
```

Example 2.4.2. Let

$$
f(z) = (\sinh(2z^2) + \sinh(10z) - 1) \times
$$
$$
(\sinh(2z^2) + \sinh(10z) - 1.01)(\sinh(2z^2) + \sinh(10z) - 1.02).
$$

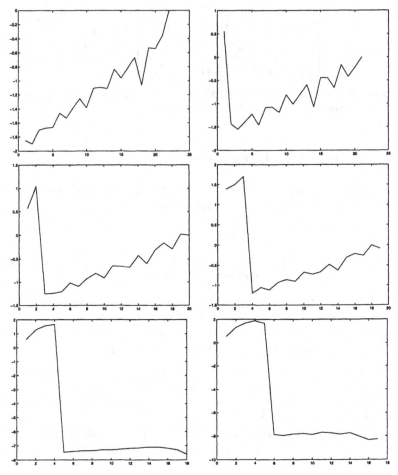

Fig. 2.1. The coefficients of $p_\sigma(z)$ for $\tau = 0, 1, \ldots, 5$

This function has 21 simple zeros inside the unit circle. They form 7 clusters, where each cluster consists of 3 zeros. Thus $N = n = 21$ and $m = 7$. This example was also studied by Sakurai et al. [111]. We set $L = 512$ and $K = 42$.

In Figure 2.2 we plot the logarithm with base 10 of the modulus of the coefficients of $p_\sigma(z)$ for $\tau = 0, 1, \ldots, 7$. The zeros of $q_7(z)$ are given by

```
-1.848537713183581e-01 - i  8.949141853554533e-01
-1.848537713183412e-01 + i  8.949141853554334e-01
-1.003354151041395e-01 - i  3.061151582728444e-01
-1.003354151030711e-01 + i  3.061151582802838e-01
 1.335489810139705e-01 - i  6.084120926164355e-01
 1.335489810131479e-01 + i  6.084120926165633e-01
 8.777826151937687e-02 + i  8.843042856595357e-12
```

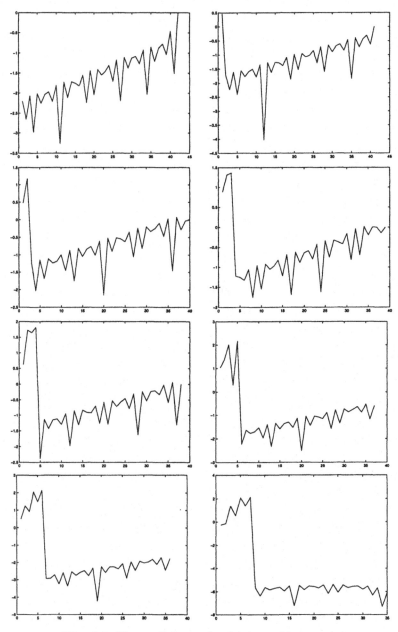

Fig. 2.2. The coefficients of $p_\sigma(z)$ for $\tau = 0, 1, \ldots, 7$

These match the approximations for the centres of the clusters that were given in [111].   ◇

3. Zeros and poles of meromorphic functions

In the previous chapters we have considered the problem of computing all the zeros of an analytic function that lie in the interior of a Jordan curve. We have seen how the algorithm that we have proposed can also be used to locate clusters of zeros. Our algorithm gives accurate results. It is based on the theory of formal orthogonal polynomials. Information concerning the location of the zeros is obtained via numerical integration.

We will show how these results can be used to tackle the problem of computing all the zeros and poles of a meromorphic function that lie in the interior of a Jordan curve. More precisely, given a meromorphic function f and a positively oriented Jordan curve γ that does not pass through any zero or pole of f, we will present an algorithm for computing all the zeros and poles of f that lie inside γ, together with their respective multiplicities and orders. An upper bound for the total number of poles of f that lie inside γ is assumed to be known. Initial approximations for the zeros and poles are not needed.

3.1 Introduction

Let W be a simply connected region in $\mathbb{C}$, $f : W \to \mathbb{C}$ analytic in W, and γ a positively oriented Jordan curve in W that does not pass through any zero of f. In Chapter 1 we have considered the problem of computing *all* the zeros $z_1, \ldots, z_n$ of f that lie inside γ, together with their respective multiplicities $\nu_1, \ldots, \nu_n$. The number of mutually distinct zeros of f that lie inside γ is denoted by n while N stands for the total number of zeros of f that lie inside γ, $N = \nu_1 + \cdots + \nu_n$. The algorithm that we have presented is based on the theory of formal orthogonal polynomials associated with the symmetric bilinear form

$$\langle \cdot, \cdot \rangle : \mathcal{P} \times \mathcal{P} \to \mathbb{C} : (\phi, \psi) \mapsto \langle \phi, \psi \rangle := \frac{1}{2\pi i} \int_\gamma \phi(z)\psi(z)\frac{f'(z)}{f(z)}\, dz.$$

As the logarithmic derivative f'/f has a simple pole at z_k with residue ν_k for $k = 1, \ldots, n$, Cauchy's Theorem implies that

$$\langle \phi, \psi \rangle = \sum_{k=1}^{n} \nu_k \phi(z_k) \psi(z_k). \tag{3.1}$$

Our algorithm proceeds as follows. It computes $z_1, \ldots, z_n$ as the zeros of the nth degree formal orthogonal polynomial (FOP) that is associated with $\langle \cdot, \cdot \rangle$. The value of n is determined indirectly. Once n and $z_1, \ldots, z_n$ have been found, the problem becomes linear and $\nu_1, \ldots, \nu_n$ are calculated by solving a Vandermonde system. The total number of zeros N can be computed via numerical integration. It is an upper bound for n. The fact that such an upper bound is available plays a crucial role in the determination of the value of n.

Now suppose that f is not analytic but meromorphic in W and suppose that f has neither zeros nor poles on γ. We will show how the algorithm that we have presented in Chapter 1 can be extended to compute *all* the zeros and poles of f that lie in the interior of γ, together with their respective multiplicities and orders. Let P denote the total number of poles of f that lie inside γ (i.e., the number of poles where each pole is counted according to its order). Suppose from now on that $N + P > 0$. Let p denote the number of mutually distinct poles of f that lie inside γ. Let $y_1, \ldots, y_p$ be these poles and $\mu_1, \ldots, \mu_p$ their respective orders. An easy calculation shows that f'/f has a simple pole at y_l with residue $-\mu_l$ for $l = 1, \ldots, p$. Therefore

$$\langle \phi, \psi \rangle = \sum_{k=1}^{n} \nu_k \phi(z_k) \psi(z_k) - \sum_{l=1}^{p} \mu_l \phi(y_l) \psi(y_l). \tag{3.2}$$

This expression is of the same type as (3.1). The theorems that we have proven in Chapter 1 do not rely on the fact that the ν_k's are known to be positive integers. They merely use the fact that $\nu_k \neq 0$ for $k = 1, \ldots, n$. It is therefore likely that these results can be extended to cover (3.2). We will see that $z_1, \ldots, z_n$ and $y_1, \ldots, y_p$ can indeed be calculated as the zeros of the FOP of degree $n + p$. Provided, of course, that an upper bound M for $m := n + p$ is known. Assume that an upper bound $\hat{P}$ for P is known and define the ordinary moments

$$s_r := \langle 1, z^r \rangle$$

for $r = 0, 1, 2, \ldots$. Then

$$s_r = \nu_1 z_1^r + \cdots + \nu_n z_n^r - \mu_1 y_1^r - \cdots - \mu_p y_p^r$$

for $r = 0, 1, 2, \ldots$. In particular, $s_0 = N - P$ and we may assume that the value of s_0 has been computed. As $n + p \leq N + P = s_0 + 2P \leq s_0 + 2\hat{P}$, it follows that we may take $M = s_0 + 2\hat{P}$. In case γ is the unit circle, an upper bound for P can be obtained by using the heuristic approach of Gleyse and Kaliaguine [58].

For previous attempts to tackle the problem of computing all the zeros and poles of a meromorphic function that lie in the interior of a Jordan curve, we refer to Abd-Elall, Delves and Reid [1] and also Ioakimidis [71, 72].

3.2 Theoretical considerations and numerical algorithm

Instead of (3.2) we will consider the following even more general setting. Let $\langle \cdot, \cdot \rangle : \mathcal{P} \times \mathcal{P} \to \mathbb{C}$ be any symmetric and bilinear form such that

$$\langle \phi, \psi \rangle = \sum_{k=1}^{m} \lambda_k \phi(x_k) \psi(x_k) \tag{3.3}$$

for any $\phi, \psi \in \mathcal{P}$, where $m \in \mathbb{N}_0$, $\lambda_1, \ldots, \lambda_m \in \mathbb{C}_0$ and $x_1, \ldots, x_m \in \mathbb{C}$ are unknown. Suppose that the points $x_1, \ldots, x_m$ are known to be mutually distinct. Assume that we have an 'oracle' at our disposal that provides us with the value of $\langle \phi, \psi \rangle$ for any $\phi, \psi \in \mathcal{P}$ upon simple request. Let an upper bound M for m be given.

Our unknowns m, $x_1, \ldots, x_m$ and $\lambda_1, \ldots, \lambda_m$ can be calculated in the same way as n, $z_1, \ldots, z_n$ and $\nu_1, \ldots, \nu_n$ have been calculated in Chapter 1. The key point is that the FOP of degree m that is associated with the form $\langle \cdot, \cdot \rangle$ is given by

$$(z - x_1) \cdots \cdot (z - x_m),$$

as one can easily verify.

Let us start by formulating what could be called the "theoretical" solution, cf. Theorems 1.2.1 and 1.2.2 and Corollary 1.3.1. This will ultimately lead to a numerical algorithm that generalizes the algorithm that we have presented in Chapter 1.

Theorems 1.2.1 and 1.2.2 have been formulated in terms of ordinary moments and Hankel matrices whereas Corollary 1.3.1 relies upon formal orthogonal polynomials. These results can be formulated in terms of polynomials from an arbitrary basis for $\mathcal{P}$. Indeed, let ψ_k be a monic polynomial of degree k for $k = 0, 1, 2, \ldots$. Define the $k \times k$ matrices F_k and $F_k^{(1)}$ as

$$F_k := \left[\langle \psi_r, \psi_s \rangle \right]_{r,s=0}^{k-1} \quad \text{and} \quad F_k^{(1)} := \left[\langle \psi_r, \psi_1 \psi_s \rangle \right]_{r,s=0}^{k-1}$$

for $k = 1, 2, \ldots$.

The following theorem characterizes m.

Theorem 3.2.1. $m = \operatorname{rank} F_{m+r}$ *for every nonnegative integer* r. *In particular,* $m = \operatorname{rank} F_M$.

Proof. Let r be a nonnegative integer. The matrix F_{m+r} can be written as

$$F_{m+r} = \sum_{k=1}^{m} \lambda_k \begin{bmatrix} \psi_0(x_k)\psi_0(x_k) & \cdots & \psi_0(x_k)\psi_{m+r-1}(x_k) \\ \vdots & & \vdots \\ \psi_{m+r-1}(x_k)\psi_0(x_k) & \cdots & \psi_{m+r-1}(x_k)\psi_{m+r-1}(x_k) \end{bmatrix}$$

$$= \sum_{k=1}^{m} \lambda_k \begin{bmatrix} \psi_0(x_k) \\ \vdots \\ \psi_{m+r-1}(x_k) \end{bmatrix} \begin{bmatrix} \psi_0(x_k) & \cdots & \psi_{m+r-1}(x_k) \end{bmatrix}.$$

This implies that rank $F_{m+r} \leq m$. However, F_m is nonsingular. Indeed, one can easily verify that F_m can be factorized as $F_m = V_m D_m V_m^T$ where V_m is the Vandermonde-like matrix

$$V_m := [\,\psi_r(x_s)\,]_{r=0,s=1}^{m-1,m}$$

and D_m is the diagonal matrix

$$D_m := \operatorname{diag}(\lambda_1, \dots, \lambda_m).$$

Therefore rank $F_{m+r} \geq m$. It follows that rank $F_{m+r} = m$.   □

Thus the regular FOP of degree m exists whereas regular FOPs of degree $> m$ do not exist. Note that, in contrast to the setting of Chapters 1 and 2, the regular FOP of degree one need not exist. This will be the only important point in which the algorithm that we will present will be different from the algorithm that we have presented in Chapter 1. If $\langle 1, 1 \rangle = \sum_{k=1}^{m} \lambda_k \neq 0$, then the regular FOP of degree one exists and it is given by $\varphi_1(z) = z - \mu$ where $\mu := \langle 1, z \rangle / \langle 1, 1 \rangle$. In the case of meromorphic functions, the condition $\langle 1, 1 \rangle \neq 0$ means that $N \neq P$. For analytic functions, we have that $P = 0$ and thus the regular FOP of degree one always exists (if we assume, of course, that $N > 0$).

The following theorem shows how $x_1, \dots, x_m$ can be computed by solving a generalized eigenvalue problem.

Theorem 3.2.2. *The eigenvalues of the pencil $F_m^{(1)} - \lambda F_m$ are given by $\psi_1(x_1), \dots, \psi_1(x_m)$.*

Proof. Define V_m as the Vandermonde-like matrix

$$V_m := [\,\psi_r(x_s)\,]_{r=0,s=1}^{m-1,m}$$

and let D_m and $D_m^{(1)}$ be the diagonal matrices

$$D_m := \operatorname{diag}(\lambda_1, \dots, \lambda_m) \quad \text{and} \quad D_m^{(1)} := \operatorname{diag}(\lambda_1 \psi_1(x_1), \dots, \lambda_m \psi_1(x_m)).$$

Then F_m and $F_m^{(1)}$ can be factorized as $F_m = V_m D_m V_m^T$ and $F_m^{(1)} = V_m D_m^{(1)} V_m^T$. Let λ^* be an eigenvalue of the pencil $F_m^{(1)} - \lambda F_m$ and x a corresponding eigenvector. Then

$$\begin{aligned}
& F_m^{(1)} x = \lambda^* F_m x \\
\Leftrightarrow\ & (V_m D_m^{(1)} V_m^T) x = \lambda^* (V_m D_m V_m^T) x \\
\Leftrightarrow\ & D_m^{(1)} y = \lambda^* D_m y \quad \text{if } y := V_m^T x \\
\Leftrightarrow\ & \operatorname{diag}(\psi_1(x_1), \dots, \psi_1(x_m)) y = \lambda^* y.
\end{aligned}$$

This proves the theorem.   □

As $\psi_1(z)$ is a polynomial of degree one, the "nodes" $x_1, \ldots, x_m$ can be calculated by applying the previous theorem. Once m and $x_1, \ldots, x_m$ have been found, the problem becomes linear and the "weights" $\lambda_1, \ldots, \lambda_m$ can be computed by solving a Vandermonde-like system of linear equations.

The question remains, of course, which polynomials $\psi_k(z)$ to choose. Again, as in Chapter 1, we will obtain very accurate numerical results by using the formal orthogonal polynomials associated with $\langle \cdot, \cdot \rangle$.

If F_m is strongly nonsingular, then we have a full set $\{\varphi_0, \varphi_1, \ldots, \varphi_m\}$ of regular FOPs. If not, then the gaps in the sequence of existing regular FOPs can be filled up in the same way as before. We may therefore assume that a suitable sequence $\{\varphi_t\}_{t=0}^{\infty}$ has been defined.

It is a trivial task to generalize Theorems 1.3.1 (concerning the computation of zeros of regular FOPs) and 1.3.2 (concerning the stopping criterium/the determination of the value of m).

These considerations lead to the following algorithm.

ALGORITHM

input $\langle \cdot, \cdot \rangle$, M, ϵ_{stop}
output m, nodes
comment nodes $= \{x_1, \ldots, x_m\}$. We assume that $M \geq m$ and $\epsilon_{\text{stop}} > 0$.

$\varphi_0(z) \leftarrow 1$
$r \leftarrow 0$
$s_0 \leftarrow \langle 1, 1 \rangle$

if $s_0 == 0$ **then**
  $\mu \leftarrow 0$
  $\varphi_1(z) \leftarrow z$
  $t \leftarrow 1$
else
  $\mu \leftarrow \langle 1, z \rangle / s_0$
  $\varphi_1(z) \leftarrow z - \mu$
  $r \leftarrow 1; t \leftarrow 0$
end if
while $r + t < M$ **do**
  regular $\leftarrow$ it is numerically feasible to generate $\varphi_{r+t+1}(z)$ as
           a regular FOP
  if regular **then**
    generate $\varphi_{r+t+1}(z)$ as a regular FOP
    $r \leftarrow r + t + 1; t \leftarrow 0$
    allsmall $\leftarrow$ **true**; $\tau \leftarrow 0$
    while allsmall **and** $(r + \tau < M)$ **do**
      $[\text{ip}, \text{maxsum}] \leftarrow \langle (z - \mu)^\tau \varphi_r(z), \varphi_r(z) \rangle$
      ip $\leftarrow |\text{ip}|$
      allsmall $\leftarrow (\text{ip}/\text{maxsum} < \epsilon_{\text{stop}})$

```
        τ ← τ + 1
    end while
    if allsmall then
        m ← r; nodes ← roots (φ_r); stop
    end if
else
    generate φ_{r+t+1}(z) as an inner polynomial
    t ← t + 1
end if
end while
m ← M; nodes ← roots (φ_N); stop
```

Similar comments as those given after the formulation of the algorithm in Chapter 1 apply.

3.3 A numerical example

We have modified the Matlab implementation that we have already used in Section 1.4 of Chapter 1.

Example 3.3.1. Suppose that

$$f(z) = \frac{1}{z^2(z-1)(z^2+9)} + z \sin z + e^{-3z} + 4$$

and let γ be the circle $\{z \in \mathbb{C} : |z| = 2\}$. We set the upper bound $\hat{P} = 5$. It turns out that $\langle 1, 1 \rangle = 0$ (in other words, $N = P$) and thus $M = s_0 + 2\hat{P} = 10$. We set $\epsilon_{\text{stop}} = 10^{-8}$. Our algorithm reacts as follows. It defines $\varphi_0(z)$ as a regular FOP, $\varphi_1(z)$ as an inner polynomial and $\varphi_2(z)$ as a regular FOP. The scaled counterpart of $|\langle \varphi_2(z), \varphi_2(z) \rangle|$ is equal to

4.821901380151357 e-01

The algorithm defines $\varphi_3(z)$ as an inner polynomial and $\varphi_4(z)$ as a regular FOP. The scaled counterpart of $|\langle \varphi_4(z), \varphi_4(z) \rangle|$ is equal to

1.776524543847086 e-01

The algorithm defines $\varphi_5(z)$ as a regular FOP. For $k = 0, 1, \ldots, 4$, the scaled counterparts of $|\langle z^k \varphi_5(z), \varphi_5(z) \rangle|$ are given by

2.387781621321191 e-15
2.887901534660305 e-15
1.268594438890019 e-15
3.904276363179922 e-15
2.772600727960076 e-15

The algorithm decides that $n + p = 5$ and stops. The absolute errors of the computed approximations for the zeros and poles are $\mathcal{O}(10^{-12})$. After one step of iterative refinement, we obtain the following results:

x_k	λ_k
0.97843635600921	1
0.16974891913248	1
−0.13327146070751	1
1.00000000000000	−1
0.00000000000000	−2

Note how zeros and poles can be distinguished by checking the signs of the λ_k's.   ◇

Note 3.3.1. As our algorithm provides not only approximations for the zeros and poles but also the corresponding multiplicities and orders, we can use the modified Newton's method

$$z_k^{(\alpha+1)} = z_k^{(\alpha)} - \nu_k \frac{f(z_k^{(\alpha)})}{f'(z_k^{(\alpha)})}, \quad y_l^{(\alpha+1)} = y_l^{(\alpha)} + \mu_l \frac{f(y_l^{(\alpha)})}{f'(y_l^{(\alpha)})}, \quad \alpha = 0, 1, 2, \ldots,$$

to refine the approximations for the zeros and poles.

4. Systems of analytic equations

In Chapter 1 we have considered the problem of computing all the zeros of an analytic function f that lie in the interior of a Jordan curve γ. The algorithm that we have presented computes not only accurate approximations for the zeros but also their respective multiplicities. It does not require initial approximations for the zeros. In Chapter 2 we have seen how our algorithm can be used to locate clusters of zeros of analytic functions whereas in Chapter 3 we have adapted it to handle the problem of calculating zeros and poles of meromorphic functions. As our approach relies upon integrals along γ that involve the logarithmic derivative f'/f, it could be called a *logarithmic residue* based approach.

In this chapter we will present a logarithmic residue based approach for the problem of computing zeros of analytic mappings (in other words, for solving systems of analytic equations). A multidimensional logarithmic residue formula is available in the literature. This formula involves the integral of a differential form, which we will transform into a sum of Riemann integrals. More precisely, if d denotes the dimension of the mapping (i.e., the number of equations, which is assumed to be equal to the number of variables), then this sum consists of d Riemann integrals of dimension $2d-1$. We will show how the zeros and their respective multiplicities can be computed from these integrals by solving a generalized eigenvalue problem that has Hankel structure and d Vandermonde systems, cf. Theorems 1.2.1 and 1.2.2 and Equation (1.19).

It turns out that these integrals are difficult to evaluate numerically. Therefore we prefer to use ordinary moments instead of modified moments, even if this means that the computed approximations for the zeros will be less accurate. By using a cubature package that can handle *vectors* of similar integrals over a common integration region, we will be able to calculate all the ordinary moments that are needed simultaneously. This will enable us to reduce the cost of our approach. Indeed, a significant part of the computation required for each integrand will be the same for all of the integrands and thus these common calculations need to be done only once for each integrand evaluation point.

4.1 Introduction

Let $d \geq 1$ be a positive integer. Consider a polydisk D in $\mathbb{C}^d$ (i.e., D is the Cartesian product of d disks in $\mathbb{C}$) and let $f = (f_1, \ldots, f_d) : \overline{D} \to \mathbb{C}^d$ be a mapping that is analytic in $\overline{D}$ and has no zeros on the boundary of D. The latter implies that f has only a finite number of zeros in D and that these zeros are all isolated [2, Theorem 2.4]. We consider the problem of computing these zeros, together with their respective multiplicities.

Let $Z_f(D)$ denote the set of zeros of f that lie in D and let $\mu_a(f)$ denote the multiplicity of a zero $a \in Z_f(D)$.

In case $d = 1$, the classical logarithmic residue formula that we have used in the previous chapters tells us that

$$\frac{1}{2\pi i} \int_{\partial D} \varphi(z) \frac{f'(z)}{f(z)} \, dz = \sum_{a \in Z_f(D)} \mu_a(f) \, \varphi(a) \tag{4.1}$$

if $\varphi : \overline{D} \to \mathbb{C}$ is analytic in D and continuous in $\overline{D}$. By evaluating the integral in the left-hand side numerically, we have been able to obtain information about the location of the zeros of f.

A multidimensional generalization of (4.1) is available in the theory of functions of several complex variables [2, Theorem 3.1]. It involves the integral of a differential form. To prepare for the numerical evaluation of this integral, we transform it into a Riemann integral, or rather, a sum of d Riemann integrals. This result is formulated in Theorem 4.2.1 and looks as follows:

$$I(\varphi) := \sum_{k=1}^{d} I_k(\varphi) = \sum_{a \in Z_f(D)} \mu_a(f) \, \varphi(a)$$

if $\varphi : \overline{D} \to \mathbb{C}$ is analytic in D and continuous in $\overline{D}$ and where $I_1(\varphi), \ldots, I_d(\varphi)$ are certain Riemann integrals over the unit cube in $\mathbb{R}^{2d-1}$. Observe that the total number of zeros of f that lie in D is equal to $I(1)$.

The proof of Theorem 4.2.1 is formulated in the language of differential forms. This cannot be avoided. Readers who feel less at ease with differential forms may consult [34], [45] or [122] for an introduction and [101] for a thorough exposition of analysis on complex manifolds.

Unfortunately, the integrals that appear in Theorem 4.2.1 tend to be difficult to evaluate numerically. The efficient numerical evaluation of these integrals is still an open problem. It represents a challenge for the numerical integration community. (More details about the various numerical integration algorithms that we have tried will be given in Section 4.4.)

Note 4.1.1. These integrals are similar to the Kronecker-Picard integrals that appear in topological degree based methods for computing solutions to systems of real equations that are twice continuously differentiable. See, for

example, Erdelsky [42], O'Neil and Thomas [102] and also Ragos, Vrahatis
and Zafiropoulos [110], and Kavvadias and Vrahatis [83].

We will therefore assume that we are able to evaluate the functional $I(\varphi)$ for
every function φ that satisfies the hypotheses of Theorem 4.2.1. In particular,
we will suppose that the total number of zeros of f that lie in D can be
computed. Our unknowns are the number of mutually distinct zeros of f
that lie in D, these zeros themselves, and their respective multiplicities. In
Section 4.3 we will show how specific choices of φ enable us to calculate
our unknowns by solving a generalized eigenvalue problem that has Hankel
structure and d Vandermonde systems.

Note 4.1.2. Algebraic mappings are of course a special case of analytic mappings. Systems of polynomial equations have received considerable interest
in recent years. Several classes of methods have been developed for their
solution: Groebner bases, homotopy continuation, sparse resultants and interval methods (see, for example, [30, 40, 43, 94, 100, 120, 121]). We will
not compare our approach with these methods because they have been developed specifically for systems of polynomial equations whereas we consider
systems of arbitrary analytic equations, a problem that has received much
less attention in the literature.

4.2 A multidimensional logarithmic residue formula

Let J_f denote the Jacobian matrix of f and let $J_{[k]}$ be the Jacobian matrix
of f with the kth column replaced with $\begin{bmatrix} f_1 & \cdots & f_d \end{bmatrix}^T$:

$$J_{[k]} := \begin{bmatrix} \frac{\partial f_1}{\partial z_1} & \cdots & \frac{\partial f_1}{\partial z_{k-1}} & f_1 & \frac{\partial f_1}{\partial z_{k+1}} & \cdots & \frac{\partial f_1}{\partial z_d} \\ \vdots & & \vdots & \vdots & \vdots & & \vdots \\ \frac{\partial f_d}{\partial z_1} & \cdots & \frac{\partial f_d}{\partial z_{k-1}} & f_d & \frac{\partial f_d}{\partial z_{k+1}} & \cdots & \frac{\partial f_d}{\partial z_d} \end{bmatrix}, \qquad k = 1, \ldots, d.$$

Suppose also that the polydisk D is given by

$$D = D_1 \times \cdots \times D_d$$

where

$$D_k = \{ z \in \mathbb{C} : |z - C_k| < R_k \}, \qquad k = 1, \ldots, d,$$

with $C_1, \ldots, C_d \in \mathbb{C}$ and $R_1, \ldots, R_d > 0$.

Theorem 4.2.1. *Let $\varphi : \overline{D} \to \mathbb{C}$ be analytic in D and continuous in $\overline{D}$.
Define $I_k(\varphi)$ for $k = 1, \ldots, d$ as the integral*

$$I_k(\varphi) := \rho_k \int_{[0,1]^{2d-1}} \frac{\varphi(z_1, \ldots, z_d) \det J_f(z_1, \ldots, z_d) \overline{\det J_{[k]}(z_1, \ldots, z_d)}}{\left(|f_1(z_1, \ldots, z_d)|^2 + \cdots + |f_d(z_1, \ldots, z_d)|^2 \right)^d}$$

$$\times e^{2\pi i \theta_k} r_1 \cdots r_{k-1} r_{k+1} \cdots r_d \, dr_1 \cdots dr_{k-1} \, dr_{k+1} \cdots dr_d \, d\theta_1 \cdots d\theta_d$$

with

$$\rho_k = \rho(d, R_1, \ldots, R_d; k) := 2^{d-1}(d-1)! \, R_1^2 \cdots R_{k-1}^2 \, R_k \, R_{k+1}^2 \cdots R_d^2$$

and where

$$z_k = z_k(\theta_k) = C_k + R_k \, e^{2\pi i \theta_k} \qquad 0 \le \theta_k \le 1$$

and

$$z_l = z_l(r_l, \theta_l) = C_l + r_l \, R_l \, e^{2\pi i \theta_l} \qquad 0 \le r_l, \theta_l \le 1$$

for $l \in \{1, \ldots, d\} \setminus \{k\}$. Then

$$I(\varphi) := \sum_{k=1}^{d} I_k(\varphi) = \sum_{a \in Z_f(D)} \mu_a(f) \, \varphi(a). \tag{4.2}$$

Theorem 4.2.1 is a corollary of the following theorem.

Theorem 4.2.2. *Let D be a polydisk in $\mathbb{C}^d$ and let $f = (f_1, \ldots, f_d) : \overline{D} \to \mathbb{C}^d$ be a mapping that is analytic in $\overline{D}$ and has no zeros on the boundary of D. Let $\varphi : \overline{D} \to \mathbb{C}$ be analytic in D and continuous in $\overline{D}$. Define the differential form $\omega(f, \overline{f})$ as*

$$\omega(f, \overline{f}) := \frac{(d-1)!}{(2\pi i)^d} \frac{1}{|f|^{2d}} \sum_{k=1}^{d} (-1)^{k-1} \overline{f}_k \, d\overline{f}_{[k]} \wedge df.$$

Then

$$\int_{\partial D} \varphi \, \omega(f, \overline{f}) = \sum_{a \in Z_f(D)} \mu_a(f) \, \varphi(a).$$

Proof. Yuzhakov and Roos [2, Theorem 3.1] proved this result for arbitrary bounded domains in $\mathbb{C}^d$ with a piecewise smooth boundary. □

The notations used in the formulation of the previous theorem,

$$\overline{f} = (\overline{f}_1, \ldots, \overline{f}_d)$$
$$|f| = \sqrt{|f_1|^2 + \cdots + |f_d|^2}$$
$$df = df_1 \wedge \cdots \wedge df_d$$
$$df_{[k]} = df_1 \wedge \cdots \wedge \widetilde{df_k} \wedge \cdots \wedge df_d, \qquad k = 1, \ldots, d,$$

are classical. (The tilde over the form df_k means that this form is omitted and does not appear in the product.)

To get rid of the differential forms in Theorem 4.2.2, we proceed as follows. Define the form $\eta(f)$ as

$$\eta(f) := \sum_{k=1}^{d}(-1)^{k-1} f_k \, df_1 \wedge \cdots \wedge \widetilde{df_k} \wedge \cdots \wedge df_d.$$

This form is sometimes called the *Leray form* [85]. Then

$$\omega(f,\overline{f}) = \frac{(d-1)!}{(2\pi i)^d} \frac{1}{|f|^{2d}} \eta(\overline{f}) \wedge df.$$

If $f = f(z_1,\ldots,z_d)$, then $df = \det J_f(z_1,\ldots,z_d) \, dz_1 \wedge \cdots \wedge dz_d$. The following lemma shows what happens with the Leray form in this case.

Lemma 4.2.1. *If $f = f(z_1,\ldots,z_d)$, then*

$$\eta(f) = \sum_{k=1}^{d}(-1)^{k-1} \det J_{[k]}(z_1,\ldots,z_d) \, dz_1 \wedge \cdots \wedge \widetilde{dz_k} \wedge \cdots \wedge dz_d.$$

Proof. Let j be an integer between 1 and d, $j \in \{1,\ldots,d\}$. Then

$$\sum_{k=1}^{d}(-1)^{k-1} \det J_{[k]} \, (dz_1 \wedge \cdots \wedge \widetilde{dz_k} \wedge \cdots \wedge dz_d)(\frac{\partial}{\partial z_1},\ldots,\frac{\widetilde{\partial}}{\partial z_j},\ldots,\frac{\partial}{\partial z_d})$$
$$= (-1)^{j-1} \det J_{[j]}.$$

By expanding $\det J_{[j]}$ along the jth column, we obtain that

$(-1)^{j-1} \det J_{[j]}$

$$= (-1)^{j-1} \sum_{k=1}^{d}(-1)^{k+j} f_k \det \begin{bmatrix} \frac{\partial f_1}{\partial z_1} & \cdots & \frac{\partial f_1}{\partial z_{j-1}} & \frac{\partial f_1}{\partial z_{j+1}} & \cdots & \frac{\partial f_1}{\partial z_d} \\ \vdots & & \vdots & \vdots & & \vdots \\ \frac{\partial f_{k-1}}{\partial z_1} & \cdots & \frac{\partial f_{k-1}}{\partial z_{j-1}} & \frac{\partial f_{k-1}}{\partial z_{j+1}} & \cdots & \frac{\partial f_{k-1}}{\partial z_d} \\ \frac{\partial f_{k+1}}{\partial z_1} & \cdots & \frac{\partial f_{k+1}}{\partial z_{j-1}} & \frac{\partial f_{k+1}}{\partial z_{j+1}} & \cdots & \frac{\partial f_{k+1}}{\partial z_d} \\ \vdots & & \vdots & \vdots & & \vdots \\ \frac{\partial f_d}{\partial z_1} & \cdots & \frac{\partial f_d}{\partial z_{j-1}} & \frac{\partial f_d}{\partial z_{j+1}} & \cdots & \frac{\partial f_d}{\partial z_d} \end{bmatrix}.$$

The determinant in the right-hand side is by definition equal to

$$(df_1 \wedge \cdots \wedge \widetilde{df_k} \wedge \cdots \wedge df_d)(\frac{\partial}{\partial z_1},\ldots,\frac{\widetilde{\partial}}{\partial z_j},\ldots,\frac{\partial}{\partial z_d}).$$

This proves the lemma. $\qquad\square$

It follows that

$$\omega(f,\overline{f}) = \sum_{k=1}^{d} \omega_k(f,\overline{f})$$

where

$$\omega_k(f,\overline{f}) := (-1)^{k-1} \frac{(d-1)!}{(2\pi i)^d} \frac{\det J_f \, \overline{\det J_{[k]}}}{|f|^{2d}} \, d\overline{z}_{[k]} \wedge dz, \qquad k = 1, \dots, d.$$

The boundary of D is given by

$$\partial D = \partial D_{[1]} \cup \cdots \cup \partial D_{[d]}$$

where

$$\partial D_{[k]} := D_1 \times \cdots \times D_{k-1} \times \partial D_k \times D_{k+1} \times \cdots \times D_d, \qquad k = 1, \dots, d,$$

and thus

$$\int_{\partial D} \varphi \omega(f,\overline{f}) = \sum_{k=1}^d \int_{\partial D_{[k]}} \varphi \omega_k(f,\overline{f}).$$

Now let us introduce polar coordinates. Let $C \in \mathbb{C}$ and $R > 0$. If $z = z(\theta) = C + R e^{2\pi i \theta}$, then $dz = 2\pi i \, R e^{2\pi i \theta} \, d\theta$. And if $z = z(r,\theta) = C + r R e^{2\pi i \theta}$, then $d\overline{z} \wedge dz = 2(2\pi i)r R^2 \, dr \wedge d\theta$. Thus

$$\int_{\partial D_{[k]}} \varphi \omega_k(f,\overline{f})$$

$$= (-1)^{k-1} \rho_k \int_{\partial D_{[k]}} \frac{\varphi \, \det J_f \, \overline{\det J_{[k]}}}{(|f_1|^2 + \cdots + |f_d|^2)^d} \Big[\prod_{l=1,\, l \neq k}^d r_l \Big] e^{2\pi i \theta_k} \, dr_{[k]} \wedge d\theta$$

for $k = 1, \dots, d$, where

$$\rho_k = \rho(d, R_1, \dots, R_d; k) := 2^{d-1}(d-1)! \, R_1^2 \cdots R_{k-1}^2 \, R_k \, R_{k+1}^2 \cdots R_d^2$$

and in the integral on the right-hand side

$$z_k = z_k(\theta_k) = C_k + R_k \, e^{2\pi i \theta_k}$$

and

$$z_l = z_l(r_l, \theta_l) = C_l + r_l \, R_l \, e^{2\pi i \theta_l}$$

for $l \in \{1, \dots, d\} \setminus \{k\}$. In [2] it is assumed that $\mathbb{C}^d$ has the orientation determined by the form $dr_1 \wedge \cdots \wedge dr_d \wedge d\theta_1 \wedge \cdots \wedge d\theta_d$ and that the boundary of D is assigned the orientation induced by that of D. This implies that $\partial D_{[k]}$ has the orientation determined by the form $(-)^{k-1} \, dr_{[k]} \wedge d\theta$. Therefore

$$\int_{\partial D_{[k]}} \varphi \omega_k(f,\overline{f}) = \rho_k$$

$$\times \int_{[0,1]^{2d-1}} \frac{\varphi(z_1, \dots, z_d) \, \det J_f(z_1, \dots, z_d) \, \overline{\det J_{[k]}(z_1, \dots, z_d)}}{(|f_1(z_1, \dots, z_d)|^2 + \cdots + |f_d(z_1, \dots, z_d)|^2)^d}$$

$$\times e^{2\pi i \theta_k} \, r_1 \cdots r_{k-1} \, r_{k+1} \cdots r_d \, dr_1 \cdots dr_{k-1} \, dr_{k+1} \cdots dr_d \, d\theta_1 \cdots d\theta_d$$

for $k = 1, \ldots, d$, where

$$z_k = z_k(\theta_k) = C_k + R_k \, e^{2\pi i \theta_k} \qquad 0 \le \theta_k \le 1$$

and

$$z_l = z_l(r_l, \theta_l) = C_l + r_l \, R_l \, e^{2\pi i \theta_l} \qquad 0 \le r_l, \theta_l \le 1$$

for $l \in \{1, \ldots, d\} \setminus \{k\}$. This proves Theorem 4.2.1.

4.3 The algorithm

In this section we will show how the zeros and their respective multiplicities can be computed by solving a generalized eigenvalue problem that has Hankel structure and d Vandermonde systems. Our approach is inspired by Theorems 1.2.1 and 1.2.2 and Equation (1.19). We have already explained why we prefer to use ordinary moments instead of modified moments.

First we introduce some notation. The total number of zeros of f that lie in D will be denoted by N. As explained in Section 4.1 we will assume that the value of N can be computed numerically. From now on, we will also suppose that $N > 0$. Let n be the number of mutually distinct zeros of f that lie in D. Let

$$(z_1^{(1)}, \ldots, z_d^{(1)}), \ldots, (z_1^{(n)}, \ldots, z_d^{(n)})$$

denote these zeros and let $\nu_1, \ldots, \nu_n$ be their respective multiplicities. Without loss of generality we may assume that $z_d^{(p)} \ne z_d^{(q)}$, $p \ne q$. Indeed, if one first applies a random unitary linear transformation to the unknowns $z = (z_1, \ldots, z_d)$, then this condition is satisfied almost surely, i.e., with probability one. Analogous results can be formulated in case $z_k^{(p)} \ne z_k^{(q)}$, $p \ne q$, for some $k \in \{1, \ldots, d-1\}$. We leave this to the reader. What happens if our algorithm is applied in case the dth components of the zeros of f that lie in D are not mutually distinct, will be discussed at the end of this section.

Define s_p for $p = 0, 1, 2, \ldots$ as

$$s_p := I(z_d^p)$$

where $I(\cdot)$ is defined in (4.2). We will assume that the sequence $(s_p)_{p \ge 0}$ can be computed numerically. By Theorem 4.2.1

$$s_p = \nu_1 \, [z_d^{(1)}]^p + \cdots + \nu_n \, [z_d^{(n)}]^p, \qquad p = 0, 1, 2, \ldots .$$

In particular, $s_0 = N$. Define

$$H_k := \begin{bmatrix} s_0 & s_1 & \ldots & s_{k-1} \\ s_1 & & \iddots & \vdots \\ \vdots & \iddots & \iddots & \vdots \\ s_{k-1} & \ldots & \ldots & s_{2k-2} \end{bmatrix}$$

for $k = 1, 2, \ldots$. Let V_n be the Vandermonde matrix with nodes $z_d^{(1)}, \ldots, z_d^{(n)}$,

$$V_n := \left[[z_d^{(l)}]^{k-1} \right]_{k,l=1}^n.$$

The following theorem characterizes n, the number of mutually distinct zeros, cf. Theorem 1.2.1.

Theorem 4.3.1. $n = \operatorname{rank} H_{n+p}$ *for every nonnegative integer p. In particular, $n = \operatorname{rank} H_N$.*

Proof. The proof is similar to that of Theorem 1.2.1. It is based on a factorization of the matrix H_n. The Vandermonde matrix V_n is regular since $z_d^{(1)}, \ldots, z_d^{(n)}$ are assumed to be mutually distinct. $\square$

The dth components $z_d^{(1)}, \ldots, z_d^{(n)}$ of the zeros can be calculated by solving a generalized eigenvalue problem that has Hankel structure, cf. Theorem 1.2.2. Define

$$H_n^< := \left[s_{1+k+l} \right]_{k,l=0}^{n-1}.$$

Theorem 4.3.2. *The eigenvalues of the pencil $H_n^< - \lambda H_n$ are given by $z_d^{(1)}, \ldots, z_d^{(n)}$.*

Proof. The proof is similar to that of Theorem 1.2.2. $\square$

Once $z_d^{(1)}, \ldots, z_d^{(n)}$ have been found, the multiplicities $\nu_1, \ldots, \nu_n$ can be computed by solving the Vandermonde system

$$V_n \begin{bmatrix} \nu_1 \\ \nu_2 \\ \vdots \\ \nu_n \end{bmatrix} = \begin{bmatrix} s_0 \\ s_1 \\ \vdots \\ s_{n-1} \end{bmatrix}. \tag{4.3}$$

Remark 4.3.1. Theoretically the $N - n$ smallest singular values of H_N are equal to zero. In practice, as by evaluating the corresponding integrals numerically we can only obtain approximations for the s_p's and because of roundoff errors in the SVD computation, this will not be the case. If the numerical rank n of H_N is difficult to determine, it is safe to consider H_N as a matrix of full rank and to solve an $N \times N$ generalized eigenvalue problem and associated Vandermonde system.

Theorem 4.3.3. *For every integer $\alpha \geq n$ the eigenvalues of the pencil*

$$\left[s_{1+k+l} \right]_{k,l=0}^{\alpha-1} - \lambda \left[s_{k+l} \right]_{k,l=0}^{\alpha-1} \tag{4.4}$$

are given by $z_d^{(1)}, \ldots, z_d^{(n)}$ and $\alpha - n$ eigenvalues that may assume arbitrary values.

Proof. This follows from Theorem 4.3.2 and by taking into account that the sequence $(s_p)_{p \geq 0}$ is a linear recurring sequence:

$$s_{n+p} + \tau_1\, s_{n-1+p} + \cdots + \tau_n\, s_p = 0, \qquad p = 0, 1, 2, \ldots,$$

if $\tau_1, \ldots, \tau_n$ are defined as the coefficients of the monic polynomial of degree n that has $z_d^{(1)}, \ldots, z_d^{(n)}$ as simple zeros,

$$\prod_{k=1}^{n} (z - z_d^{(k)}) =: z^n + \tau_1\, z^{n-1} + \cdots + \tau_n.$$

Note that the latter implies that the matrices

$$\left[s_{1+k+l} \right]_{k,l=0}^{\alpha-1} \qquad \text{and} \qquad \left[s_{k+l} \right]_{k,l=0}^{\alpha-1}$$

have the same null spaces. $\qquad\qquad\qquad\qquad\qquad\qquad\qquad\qquad\qquad\qquad\square$

The previous theorem tells us that the generalized eigenvalue problem (4.4) has $\alpha - n$ eigenvalues that may assume arbitrary values. Each of these indeterminate eigenvalues corresponds to two corresponding zeros on the diagonals of the generalized Schur decomposition of the Hankel matrices $\left[s_{1+k+l} \right]_{k,l=0}^{\alpha-1}$ and $\left[s_{k+l} \right]_{k,l=0}^{\alpha-1}$. When actually calculated, these diagonal entries are different from zero because of roundoff errors. The quotient of two such corresponding diagonal entries is an eigenvalue that is not the dth component of a zero of f. Fortunately, cf. the concluding paragraph of Example 1.4.3, the corresponding Vandermonde system enables us to detect such spurious "dth components." Indeed, the Vandermonde matrix whose nodes are given by the eigenvalues of (4.4) is still regular and therefore the corresponding Vandermonde system has only one solution, which gives the true dth components their correct corresponding multiplicity and the spurious ones "multiplicity" zero.

Once n, $z_d^{(1)}, \ldots, z_d^{(n)}$ and $\nu_1, \ldots, \nu_n$ have been found, the unknowns

$$z_1^{(1)}, \ldots, z_1^{(n)}, \ldots, z_{d-1}^{(1)}, \ldots, z_{d-1}^{(n)}$$

can be obtained as follows. Define $t_{k,p}$ for $k = 1, \ldots, d-1$ and $p = 0, 1, 2, \ldots$ as

$$t_{k,p} := I(z_k\, z_d^p).$$

We will assume that the sequences $(t_{k,p})_{p \geq 0}$ can be computed for $k = 1, \ldots, d-1$. By Theorem 4.2.1

$$t_{k,p} = \nu_1\, z_k^{(1)}\, [z_d^{(1)}]^p + \cdots + \nu_n\, z_k^{(n)}\, [z_d^{(n)}]^p$$

for $k = 1, \ldots, d-1$ and $p = 0, 1, 2, \ldots$. It follows that $z_k^{(1)}, \ldots, z_k^{(n)}$ can be calculated from the solution of the Vandermonde system

$$V_n \begin{bmatrix} \nu_1 \, z_k^{(1)} \\ \nu_2 \, z_k^{(2)} \\ \vdots \\ \nu_n \, z_k^{(n)} \end{bmatrix} = \begin{bmatrix} t_{k,0} \\ t_{k,1} \\ \vdots \\ t_{k,n-1} \end{bmatrix}, \qquad k = 1, \ldots, d-1.$$

As already discussed in Chapter 1, problems of numerical linear algebra that involve Vandermonde or Hankel matrices truly deserve their reputation of being ill-conditioned [55, 117]. By moving the origin in the z_d-plane to the arithmetic mean of the dth components of the zeros,

$$\acute{z}_d := \frac{\nu_1 \, z_d^{(1)} + \cdots + \nu_n \, z_d^{(n)}}{\nu_1 + \cdots + \nu_n} = \frac{I(z_d)}{I(1)},$$

ill-conditioning is reduced significantly. Therefore we will use shifted versions of the integrals s_p and $t_{k,p}$, denoted by $\hat{s}_p$ and $\hat{t}_{k,p}$. The results of this section then lead to the following algorithm.

ALGORITHM

1. $N \leftarrow I(1)$
2. $\acute{z}_d \leftarrow I(z_d)/N$
3. $\hat{s}_0 \leftarrow N; \quad \hat{s}_1 \leftarrow 0; \quad \hat{s}_p \leftarrow I((z_d - \acute{z}_d)^p)$ for $p = 2, \ldots, 2N-1$
4. $n \leftarrow \mathrm{rank} \left[\hat{s}_{k+l} \right]_{k,l=0}^{N-1}$
5. Calculate the eigenvalues $\lambda_1, \ldots, \lambda_n$ of the pencil

$$\left[\hat{s}_{1+k+l} \right]_{k,l=0}^{n-1} - \lambda \left[\hat{s}_{k+l} \right]_{k,l=0}^{n-1}.$$

6. $z_d^{(k)} \leftarrow \lambda_k + \acute{z}_d$ for $k = 1, \ldots, n$
7. Solve the Vandermonde system

$$\begin{bmatrix} 1 & 1 & \cdots & 1 \\ \lambda_1 & \lambda_2 & \cdots & \lambda_n \\ \vdots & \vdots & & \vdots \\ \lambda_1^{n-1} & \lambda_2^{n-1} & \cdots & \lambda_n^{n-1} \end{bmatrix} \begin{bmatrix} \nu_1 \\ \nu_2 \\ \vdots \\ \nu_n \end{bmatrix} = \begin{bmatrix} \hat{s}_0 \\ \hat{s}_1 \\ \vdots \\ \hat{s}_{n-1} \end{bmatrix}.$$

8. $\hat{t}_{k,p} \leftarrow I(z_k(z_d - \acute{z}_d)^p)$ for $k = 1, \ldots, d-1$ and $p = 0, \ldots, n-1$
9. Solve the Vandermonde system

$$\begin{bmatrix} 1 & 1 & \cdots & 1 \\ \lambda_1 & \lambda_2 & \cdots & \lambda_n \\ \vdots & \vdots & & \vdots \\ \lambda_1^{n-1} & \lambda_2^{n-1} & \cdots & \lambda_n^{n-1} \end{bmatrix} \begin{bmatrix} x_{1,1} & \cdots & x_{d-1,1} \\ x_{1,2} & \cdots & x_{d-1,2} \\ \vdots & & \vdots \\ x_{1,n} & \cdots & x_{d-1,n} \end{bmatrix} = \begin{bmatrix} \hat{t}_{1,0} & \cdots & \hat{t}_{d-1,0} \\ \hat{t}_{1,1} & \cdots & \hat{t}_{d-1,1} \\ \vdots & & \vdots \\ \hat{t}_{1,n-1} & \cdots & \hat{t}_{d-1,n-1} \end{bmatrix}.$$

$z_k^{(l)} \leftarrow x_{k,l}/\nu_l$ for $k = 1, \ldots, d-1$ and $l = 1, \ldots, n$

In practice, as explained in the previous remark, we may dispense with step 4 and solve an $N \times N$ generalized eigenvalue problem and associated Vandermonde system. The components of the solution of the latter will be rounded to the nearest integers and eigenvalues that have "multiplicity" zero are thrown away.

Remark 4.3.2. What happens if our algorithm is applied in case the dth components of the zeros of f that lie in D are not mutually distinct? In this case only a subset of the zeros will be determined correctly, namely the zeros whose dth component occurs only once. To illustrate what happens to the other zeros, suppose for example that $z_d^{(1)} = z_d^{(2)}$ whereas $z_d^{(3)}, \ldots, z_d^{(n)}$ are mutually distinct and

$$\{ z_d^{(1)} \} \cap \{ z_d^{(3)}, \ldots, z_d^{(n)} \} = \emptyset.$$

Then

$$s_p = (\nu_1 + \nu_2) [z_d^{(1)}]^p + \nu_3 [z_d^{(3)}]^p + \cdots + \nu_n [z_d^{(n)}]^p$$

for $p = 0, 1, 2, \ldots$ and

$$t_{k,p} = (\nu_1 z_k^{(1)} + \nu_2 z_k^{(2)}) [z_d^{(1)}]^p + \nu_3 z_k^{(3)} [z_d^{(3)}]^p + \cdots + \nu_n z_k^{(n)} [z_d^{(n)}]^p$$

for $k = 1, \ldots, d - 1$ and $p = 0, 1, 2, \ldots$. It follows that our algorithm will replace each group of zeros that have the same dth component by a point in $\mathbb{C}^d$ whose dth component is equal to the dth component that is shared by these zeros and whose other components are given by the arithmetic mean of the corresponding components of these zeros.

4.4 Numerical examples

Let us discuss a few numerical examples.

Example 4.4.1. We have chosen $d = 4$ (the number of variables) and $n = 5$ (the number of mutually distinct zeros). The zeros and their respective multiplicities are listed in the following table.

$(z_1^{(k)}, \ldots, z_d^{(k)})$	ν_k
$(\ 1,\ \ i, -2,\ 1)$	1
$(-3,\ \ 0,\ \ 2,\ 2)$	2
$(\ 2, -i,\ \ 5,\ 3)$	1
$(\ 1,\ \ 0, -1,\ 4)$	2
$(\ 0,\ \ 1,\ \ 3,\ 5)$	1

Then $N = 7$ (the total number of zeros) and $z_d' = (1 \cdot 1 + 2 \cdot 2 + 1 \cdot 3 + 2 \cdot 4 + 1 \cdot 5)/7 = 3$ (the arithmetic mean of the last components). The shifted moments for the last components are given by

$$\hat{s}_p = 1 \cdot (1-3)^p + 2 \cdot (2-3)^p + 1 \cdot (3-3)^p + 2 \cdot (4-3)^p + 1 \cdot (5-3)^p$$

and thus $\hat{s}_p = (2+2^p)(1+(-1)^p)$ for $p \geq 1$. We computed the eigenvalues of the pencil

$$\left[\hat{s}_{1+k+l}\right]_{k,l=0}^{N-1} - \lambda \left[\hat{s}_{k+l}\right]_{k,l=0}^{N-1}$$

using LAPACK's routine CGEGV [8]. The calculations were performed in single precision arithmetic on an IBM SP2. The results are shown in the following table, in which each generalized eigenvalue is represented as a pair (α_k, β_k) of two corresponding diagonal entries in the generalized Schur decomposition of $\left[\hat{s}_{1+k+l}\right]_{k,l=0}^{N-1}$ and $\left[\hat{s}_{k+l}\right]_{k,l=0}^{N-1}$.

α_k	β_k	$z_d^{(k)}$	ν_k
-30316.83	15158.42	1.000000	1.000001
-2322.758	2322.756	1.999999	1.999996
0.003323	542.2963	3.000006	1.000017
950.4792	950.4736	4.000006	1.999985
33612.10	16806.05	5.000000	0.999998
0.000000	0.000233		
0.000000	0.000102		

The number of mutually distinct zeros is clearly equal to five. The generalized eigenvalues α_k/β_k corresponding to the last two pairs (α_k, β_k) represent spurious dth components and are thrown away. Also shown in this table are the computed values $z_d + \alpha_k/\beta_k$ of the dth components and the corresponding solutions of the Vandermonde system (4.3) for the multiplicities. The computed approximations for the other components of the zeros are shown in the following table.

$z_1^{(k)}$	$z_2^{(k)}$	$z_3^{(k)}$
$1.000002 - i\,0.000000$	$0.000000 + i\,1.000001$	$-2.000006 - i\,0.000000$
$-2.999997 + i\,0.000000$	$-0.000000 - i\,0.000002$	$2.000013 + i\,0.000001$
$2.000010 - i\,0.000000$	$0.000000 - i\,1.000002$	$4.999987 - i\,0.000002$
$0.999992 + i\,0.000000$	$-0.000001 + i\,0.000002$	$-1.000007 + i\,0.000001$
$-0.000002 + i\,0.000000$	$1.000000 - i\,0.000000$	$3.000006 - i\,0.000000$

The previous example illustrates how our algorithm is able to compute the unknown zeros and multiplicities from the integrals $\hat{s}_p$ and $\hat{t}_{k,p}$. As already mentioned in Section 4.1, the efficient numerical evaluation of these integrals is a problem that remains to be tackled. It represents a challenge for the numerical integration community. Nevertheless we would like to present a small example in which these integrals have been calculated numerically.

Example 4.4.2. Consider the problem of computing all the zeros of

$$f = f(z_1, z_2) = (\sin z_1 + z_1^2 + e^{z_2} - \cos(2z_2), \cos z_1 + z_2^3 + e^{2z_2} - 2)$$

that lie in the polydisk

$$D = \{z_1 \in \mathbb{C} : |z_1| < 1\} \times \{z_2 \in \mathbb{C} : |z_2| < 1\}.$$

In this case we have to integrate over the unit cube in $\mathbb{R}^3$. We tried several numerical integration strategies:

- lattice rules (see Joe and Sloan [79, 80, 113], Beckers and Haegemans [14], Sidi [112] and Laurie [92]);
- Monte Carlo methods (see, e.g., Kalos and Whitlock [82] and Fishman [44]);
- the software package DCUHRE [15], written by Berntsen, Espelid and Genz.

DCUHRE gave the best results. This package implements an adaptive algorithm for numerical integration over hyperrectangular regions. First we calculated $\operatorname{Re} I(1)$. We requested an absolute accuracy of 0.1 and obtained that $\operatorname{Re} I(1) \approx 1.998$. Thus $N = 2$. Next we calculated the arithmetic mean of the second components. A crude approximation for this mean is sufficient to reduce ill-conditioning and therefore we requested a relative accuracy of only 0.1. Finally we calculated all the other integrals needed by our algorithm. One of the very interesting features of DCUHRE is that it is able to integrate a *vector* of similar integrals over a common integration region. Since a significant part of the computation required for each integrand is the same for all of the integrands, these common calculations need to be done only once for each integrand evaluation point. We requested a relative accuracy of 10^{-5}. DCUHRE needed about 10^5 functional evaluations to obtain this accuracy. With these approximations for the integrals $\hat{s}_p$ and $\hat{t}_{k,p}$ as input, our algorithm obtained that f has two zeros in D ($n = 2$), each of multiplicity one. We refined the approximations for these zeros iteratively via Newton's method. The zeros of f that lie in D are given by $(0,0)$ and $(-0.72011062161456, 0.11033979708375)$. $\Diamond$

References

1. L. F. Abd-Elall, L. M. Delves, and J. D. Reid, *A numerical method for locating the zeros and poles of a meromorphic function*, Numerical Methods for Nonlinear Algebraic Equations (P. Rabinowitz, ed.), Gordon and Breach, 1970, pp. 47–59.
2. I. A. Aĭzenberg and A. P. Yuzhakov, *Integral representations and residues in multidimensional complex analysis*, Translations of Mathematical Monographs, vol. 58, American Mathematical Society, Providence, Rhode Island, 1983.
3. E. G. Anastasselou, *A formal comparison of the Delves-Lyness and Burniston-Siewert methods for locating the zeros of analytic functions*, IMA J. Numer. Anal. **6** (1986), 337–341.
4. E. G. Anastasselou and N. I. Ioakimidis, *Application of the Cauchy theorem to the location of zeros of sectionally analytic functions*, J. Appl. Math. Phys. **35** (1984), 705–711.
5. _____, *A generalization of the Siewert-Burniston method for the determination of zeros of analytic functions*, J. Math. Phys. **25** (1984), no. 8, 2422–2425.
6. _____, *A new method for obtaining exact analytical formulae for the roots of transcendental functions*, Lett. Math. Phys. **8** (1984), 135–143.
7. _____, *A new approach to the derivation of exact analytical formulae for the zeros of sectionally analytic functions*, J. Math. Anal. Appl. **112** (1985), no. 1, 104–109.
8. E. Anderson, Z. Bai, C. Bischof, J. Demmel, J. Dongarra, J. Du Croz, A. Greenbaum, S. Hammarling, A. McKenney, S. Ostrouchov, and D. Sorensen, *LAPACK users' guide*, SIAM, 1994.
9. A. C. Antoulas, *On the scalar rational interpolation problem*, IMA J. Math. Control Inf. **3** (1986), 61–88.
10. _____, *Rational interpolation and the Euclidean algorithm*, Linear Algebr. Appl. **108** (1988), 157–171.
11. A. C. Antoulas, J. A. Ball, J. Kang, and J. C. Willems, *On the solution of the minimal rational interpolation problem*, Linear Algebr. Appl. **137/138** (1990), 511–573.
12. L. Atanassova, *On the simultaneous determination of the zeros of an analytic function inside a simple smooth closed contour in the complex plane*, J. Comput. Appl. Math. **50** (1994), 99–107.
13. B. Beckermann and E. Bourreau, *How to choose modified moments?*, J. Comput. Appl. Math. **98** (1998), 81–98.
14. M. Beckers and A. Haegemans, *Transformations of integrands for lattice rules*, Numerical Integration: Recent Developments, Software and Applications (T. O. Espelid and A. Genz, eds.), Kluwer Academic Publishers, 1992, pp. 329–340.
15. J. Berntsen, T. O. Espelid, and A. Genz, *Algorithm 698: DCUHRE—An adaptive multidimensional integration routine for a vector of integrals*, ACM Trans. Math. Softw. **17** (1991), no. 4, 452–456.

16. J.-P. Berrut and H. D. Mittelmann, *Matrices for the direct determination of the barycentric weights of rational interpolation*, J. Comput. Appl. Math. **78** (1997), no. 2, 355–370.

17. D. A. Bini, *Numerical computation of polynomial zeros by means of Aberth's method*, Numerical Algorithms **13** (1996), 179–200.

18. D. A. Bini, L. Gemignani, and B. Meini, *Factorization of analytic functions by means of Koenig's Theorem and Toeplitz computations*, Unpublished manuscript.

19. D. A. Bini and V. Y. Pan, *Graeffe's, Chebyshev-like, and Cardinal's processes for splitting a polynomial into factors*, J. Complexity **12** (1996), 492–511.

20. A. W. Bojanczyk and G. Heinig, *A multi-step algorithm for Hankel matrices*, J. Complexity **10** (1994), no. 1, 142–164.

21. L. C. Botten, M. S. Craig, and R. C. McPhedran, *Complex zeros of analytic functions*, Comput. Phys. Commun. **29** (1983), no. 3, 245–259.

22. T. Boult and K. Sikorski, *An optimal complexity algorithm for computing the topological degree in two dimensions*, SIAM J. Sci. Stat. Comput. **10** (1989), no. 4, 686–698.

23. A. G. Buckley, *Conversion of Fortran 90: a case study*, ACM Trans. Math. Software **20** (1994), no. 3, 308–353.

24. A. Bultheel and M. Van Barel, *Euclid, Padé and Lanczos: another golden braid*, Report TW 188, Department of Computer Science, K.U.Leuven, April 1993.

25. _____, *Formal orthogonal polynomials for arbitrary moment matrix and Lanczos type methods*, Proceedings of the Cornelius Lanczos International Centenary Conference (Philadelphia, PA) (J. D. Brown, M. T. Chu, D. C. Ellison, and R. J. Plemmons, eds.), Society for Industrial and Applied Mathematics, 1994, pp. 273–275.

26. _____, *Linear algebra, rational approximation and orthogonal polynomials*, Studies in Computational Mathematics, vol. 6, North-Holland, 1997.

27. E. E. Burniston and C. E. Siewert, *The use of Riemann problems in solving a class of transcendental equations*, Proc. Camb. Philos. Soc. **73** (1973), 111–118.

28. S. Cabay and R. Meleshko, *A weakly stable algorithm for Padé approximants and the inversion of Hankel matrices*, SIAM J. Matrix Anal. Appl. **14** (1993), no. 3, 735–765.

29. J.-P. Cardinal, *On two iterative methods for approximating the roots of a polynomial*, The Mathematics of Numerical Analysis (Providence, Rhode Island) (J. Renegar, M. Shub, and S. Smale, eds.), Lectures in Applied Mathematics, vol. 32, 1995 AMS-SIAM Summer Seminar in Applied Mathematics, July 17–August 11, 1995, Park City, Utah, American Mathematical Society, 1996, pp. 165–188.

30. J.-P. Cardinal and B. Mourrain, *Algebraic approach of residues and applications*, The Mathematics of Numerical Analysis (Providence, Rhode Island) (J. Renegar, M. Shub, and S. Smale, eds.), Lectures in Applied Mathematics, vol. 32, 1995 AMS-SIAM Summer Seminar in Applied Mathematics, July 17–August 11, 1995, Park City, Utah, American Mathematical Society, 1996, pp. 189–210.

31. M. P. Carpentier and A. F. Dos Santos, *Solution of equations involving analytic functions*, J. Comput. Phys. **45** (1982), 210–220.

32. C. Carstensen and T. Sakurai, *Simultaneous factorization of a polynomial by rational approximation*, J. Comput. Appl. Math. **61** (1995), no. 2, 165–178.

33. A. Córdova, W. Gautschi, and S. Ruscheweyh, *Vandermonde matrices on the circle: spectral properties and conditioning*, Numer. Math. **57** (1990), no. 6/7, 577–591.

34. M. Crampin and F. A. E. Pirani, *Applicable differential geometry*, London Mathematical Society Lecture Note Series, vol. 59, Cambridge University Press, Cambridge, 1986.

35. B. Davies, *Locating the zeros of an analytic function*, J. Comput. Phys. **66** (1986), 36–49.

36. L. M. Delves and J. N. Lyness, *A numerical method for locating the zeros of an analytic function*, Math. Comput. **21** (1967), 543–560.

37. A. Draux, *Polynômes orthogonaux formels—applications*, Lecture Notes in Mathematics, vol. 974, Springer, 1983.

38. _____, *Formal orthogonal polynomials revisited. Applications*, Numer. Algorithms **11** (1996), 143–158.

39. Ö. Eğecioğlu and Ç. K. Koç, *A fast algorithm for rational interpolation via orthogonal polynomials*, Math. Comput. **53** (1989), no. 187, 249–264.

40. I. Z. Emiris and J. F. Canny, *Efficient incremental algorithms for the sparse resultant and the mixed volume*, J. Symbolic Computation **20** (1995), no. 2, 117–149.

41. K. Engelborghs, T. Luzyanina, and D. Roose, *Bifurcation analysis of periodic solutions of neutral functional differential equations: A case study*, Int. J. Bifurcation Chaos **8** (1998), no. 10, 1889–1905.

42. P. J. Erdelsky, *Computing the Brouwer degree in* $\mathbb{R}^2$, Math. Comput. **27** (1973), no. 121, 133–137.

43. J. C. Faugère, P. Gianni, D. Lazard, and T. Mora, *Efficient computation of zero-dimensional Gröbner bases by change of ordering*, J. Symbolic Computation **16** (1993), no. 4, 329–344.

44. G. S. Fishman, *Monte Carlo: Concepts, algorithms, and applications*, Springer, 1996.

45. H. Flanders, *Differential forms with applications to the physical sciences*, Dover, New York, 1989.

46. R. W. Freund and H. Zha, *A look-ahead algorithm for the solution of general Hankel systems*, Numer. Math. **64** (1993), 295–321.

47. F. D. Gakhov, *Boundary value problems*, International Series of Monographs in Pure and Applied Mathematics, vol. 85, Pergamon Press, Oxford, 1966.

48. W. Gautschi, *On inverses of Vandermonde and confluent Vandermonde matrices*, Numer. Math. **4** (1962), 117–123.

49. _____, *On inverses of Vandermonde and confluent Vandermonde matrices. II*, Numer. Math. **5** (1963), 425–430.

50. _____, *On the construction of Gaussian quadrature rules from modified moments*, Math. Comput. **24** (1970), no. 110, 245–260.

51. _____, *Optimally conditioned Vandermonde matrices*, Numer. Math. **24** (1975), no. 1, 1–12.

52. _____, *On inverses of Vandermonde and confluent Vandermonde matrices. III*, Numer. Math. **29** (1978), 445–450.

53. _____, *On generating orthogonal polynomials*, SIAM J. Sci. Stat. Comput. **3** (1982), no. 3, 289–317.

54. _____, *On the sensitivity of orthogonal polynomials to perturbations in the moments*, Numer. Math. **48** (1986), 369–382.

55. _____, *How (un)stable are Vandermonde systems?*, Asymptotic and Computational Analysis: Conference in Honor of Frank W. J. Olver's 65th Birthday (R. Wong, ed.), Lecture Notes in Pure and Applied Mathematics, vol. 124, Marcel Dekker, 1990, pp. 193–210.

56. W. Gautschi and G. Inglese, *Lower bounds for the condition number of Vandermonde matrices*, Numer. Math. **52** (1988), 241–250.

57. L. Gemignani, *Rational interpolation via orthogonal polynomials*, Computers Math. Applic. **26** (1993), no. 5, 27–34.

58. B. Gleyse and V. Kaliaguine, *On algebraic computation of number of poles of meromorphic functions in the unit disk*, Nonlinear Numerical Methods and Rational Approximation II (A. Cuyt, ed.), Kluwer Academic Publishers, 1994, pp. 241–246.

59. I. Gohberg and I. Koltracht, *Mixed, componentwise, and structured condition numbers*, SIAM J. Matrix Anal. Appl. **14** (1993), no. 3, 688–704.

60. G. H. Golub, P. Milanfar, and J. Varah, *A stable numerical method for inverting shape from moments*, Unpublished manuscript.

61. W. B. Gragg and M. H. Gutknecht, *Stable look-ahead versions of the Euclidean and Chebyshev algorithms*, Approximation and Computation: A Festschrift in Honor of Walter Gautschi (R. V. M. Zahar, ed.), Birkhäuser, 1994, pp. 231–260.

62. M. H. Gutknecht, *Continued fractions associated with the Newton–Padé table*, Numer. Math. **56** (1989), 547–589.

63. _____, *In what sense is the rational interpolation problem well posed?*, Constr. Approx. **6** (1990), no. 4, 437–450.

64. _____, *The rational interpolation problem revisited*, Rocky Mt. J. Math. **21** (1991), no. 1, 263–280.

65. _____, *Block structure and recursiveness in rational interpolation*, Approximation Theory VII (E. W. Cheney, C. K. Chui, and L. L. Schumaker, eds.), Academic Press, 1992, pp. 93–130.

66. _____, *A completed theory of the unsymmetric Lanczos process and related algorithms, Part I*, SIAM J. Matrix Anal. Appl. **13** (1992), no. 2, 594–639.

67. _____, *A completed theory of the unsymmetric Lanczos process and related algorithms, Part II*, SIAM J. Matrix Anal. Appl. **15** (1994), no. 1, 15–58.

68. P. Henrici, *Applied and computational complex analysis: I. Power series—integration—conformal mapping—location of zeros*, Wiley, 1974.

69. J. Herlocker and J. Ely, *An automatic and guaranteed determination of the number of roots of an analytic function interior to a simple closed curve in the complex plane*, Reliable Computing **1** (1995), no. 3, 239–250.

70. V. Hribernig and H. J. Stetter, *Detection and validation of clusters of polynomial zeros*, J. Symbolic Computation **24** (1997), no. 6, 667–681.

71. N. I. Ioakimidis, *Application of the generalized Siewert-Burniston method to locating zeros and poles of meromorphic functions*, J. Appl. Math. Phys. **36** (1985), 733–742.

72. _____, *Determination of poles of sectionally meromorphic functions*, J. Comput. Appl. Math. **15** (1986), 323–327.

73. _____, *Quadrature methods for the determination of zeros of transcendental functions—a review*, Numerical Integration: Recent Developments, Software and Applications (P. Keast and G. Fairweather, eds.), Reidel, Dordrecht, The Netherlands, 1987, pp. 61–82.

74. _____, *A unified Riemann-Hilbert approach to the analytical determination of zeros of sectionally analytic functions*, J. Math. Anal. Appl. **129** (1988), no. 1, 134–141.

75. _____, *A note on the closed-form determination of zeros and poles of generalized analytic functions*, Stud. Appl. Math. **81** (1989), 265–269.

76. N. I. Ioakimidis and E. G. Anastasselou, *A modification of the Delves-Lyness method for locating the zeros of analytic functions*, J. Comput. Phys. **59** (1985), 490–492.

77. _____, *A new, simple approach to the derivation of exact analytical formulae for the zeros of analytic functions*, Appl. Math. Comput. **17** (1985), 123–127.

78. _____, *On the simultaneous determination of zeros of analytic or sectionally analytic functions*, Computing **36** (1986), 239–247.
79. S. Joe and I. H. Sloan, *Implementation of a lattice method for numerical multiple integration*, ACM Trans. Math. Softw. **19** (1993), 523–545.
80. _____, *Corrigendum*, ACM Trans. Math. Softw. **20** (1994), no. 2, 245.
81. E. Jonckheere and C. Ma, *A simple Hankel interpretation of the Berlekamp-Massey algorithm*, Linear Algebr. Appl. **125** (1989), 65–76.
82. M. H. Kalos and P. A. Whitlock, *Monte Carlo methods. Volume I: Basics*, Wiley, 1986.
83. D. J. Kavvadias and M. N. Vrahatis, *Locating and computing all the simple roots and extrema of a function*, SIAM J. Sci. Comput. **17** (1996), no. 5, 1232–1248.
84. P. Kirrinnis, *Newton iteration towards a cluster of polynomial zeros*, Foundations of Computational Mathematics (F. Cucker and M. Shub, eds.), Springer, 1997, pp. 193–215.
85. S. G. Krantz, *Function theory of several complex variables*, Wiley, 1982.
86. P. Kravanja, M. Van Barel, and A. Haegemans, *Logarithmic residue based methods for computing zeros of analytic functions and related problems*, HERCMA '98: Proceedings of the Fourth Hellenic-European Conference on Computer Mathematics and its Applications (E. A. Lipitakis, ed.), September 24–26, 1998 - Athens, Hellas, LEA, 1999, pp. 201–208.
87. P. Kravanja, R. Cools, and A. Haegemans, *Computing zeros of analytic mappings: A logarithmic residue approach*, BIT **38** (1998), no. 3, 583–596, [Zbl 916.65053].
88. P. Kravanja, T. Sakurai, and M. Van Barel, *On locating clusters of zeros of analytic functions*, BIT **39** (1999), no. 4, 646–682.
89. P. Kravanja and M. Van Barel, *A derivative-free algorithm for computing zeros of analytic functions*, Computing **63** (1999), no. 1, 69–91.
90. P. Kravanja, M. Van Barel, and A. Haegemans, *On computing zeros and poles of meromorphic functions*, Computational Methods and Function Theory 1997 (N. Papamichael, St. Ruscheweyh, and E. B. Saff, eds.), Series in Approximations and Decompositions, vol. 11, World Scientific, 1999, Proceedings of the third CMFT conference, 13–17 October 1997, Nicosia, Cyprus, pp. 359–369.
91. P. Kravanja, M. Van Barel, O. Ragos, M. N. Vrahatis, and F. A. Zafiropoulos, *ZEAL: A mathematical software package for computing zeros of analytic functions*, Comput. Phys. Commun. **124** (2000), no. 2–3, 212–232.
92. D. P. Laurie, *Periodizing transformations for numerical integration*, J. Comput. Appl. Math. **66** (1996), no. 1–2, 337–344.
93. T.-Y. Li, *On locating all zeros of an analytic function within a bounded domain by a revised Delves/Lyness method*, SIAM J. Numer. Anal. **20** (1983), no. 4, 865–871.
94. _____, *Numerical solutions of multivariate polynomial systems by homotopy continuation methods*, Acta Numerica, vol. 6, Cambridge University Press, 1997, pp. 399–436.
95. N. G. Lloyd, *Degree theory*, Cambridge Tracts in Mathematics, vol. 73, Cambridge University Press, 1978.
96. J. N. Lyness and L. M. Delves, *On numerical contour integration round a closed contour*, Math. Comput. **21** (1967), 561–577.
97. J. M. McNamee, *A bibliography on roots of polynomials*, J. Comput. Appl. Math. **47** (1993), 391–394.
98. _____, *A supplementary bibliography on roots of polynomials*, J. Comput. Appl. Math. **78** (1997), no. 3, 1.
99. J. Meinguet, *On the solubility of the Cauchy interpolation problem*, Approximation Theory (A. Talbot, ed.), Academic Press, 1970, pp. 137–163.

100. B. Mourrain and V. Y. Pan, *Solving special polynomial systems by using structured matrices and algebraic residues*, Foundations of Computational Mathematics (F. Cucker and M. Shub, eds.), Springer, 1997, pp. 287–304.
101. R. Narasimhan, *Analysis on real and complex manifolds*, Advanced Studies in Pure Mathematics, vol. 1, North-Holland, Amsterdam, 1968.
102. T. O'Neil and J. W. Thomas, *The calculation of the topological degree by quadrature*, SIAM J. Numer. Anal. **12** (1975), no. 5, 673–680.
103. V. Y. Pan, *Solving a polynomial equation: some history and recent progress*, SIAM Review **39** (1997), no. 2, 187–220.
104. J. R. Partington, *An introduction to Hankel operators*, London Mathematical Society Student Texts, vol. 13, Cambridge University Press, 1988.
105. M. S. Petković, *Inclusion methods for the zeros of analytic functions*, Computer Arithmetic and Enclosure Methods (L. Átanassova and J. Herzberger, eds.), North-Holland, 1992, pp. 319–328.
106. M. S. Petković, C. Carstensen, and M. Trajković, *Weierstrass formula and zero-finding methods*, Numer. Math. **69** (1995), 353–372.
107. M. S. Petković and D. Herceg, *Higher-order iterative methods for approximating zeros of analytic functions*, J. Comput. Appl. Math. **39** (1992), 243–258.
108. M. S. Petković and Z. M. Marjanović, *A class of simultaneous methods for the zeros of analytic functions*, Comput. Math. Appl. **22** (1991), no. 10, 79–87.
109. R. Piessens, E. de Doncker-Kapenga, C. W. Überhuber, and D. K. Kahaner, *QUADPACK: A subroutine package for automatic integration*, Springer Series in Computational Mathematics, vol. 1, Springer, 1983.
110. O. Ragos, M. N. Vrahatis, and F. A. Zafiropoulos, *The topological degree for the computation of the exact number of equilibrium points of dynamical systems*, Hellenic European Research on Mathematics and Informatics '94 (HERMIS '94). Proceedings of the second Hellenic European Conference on Mathematics and Informatics, Athens (Greece), September 22–24, 1994 (E. A. Lipitakis, ed.), 1994, pp. 533–542.
111. T. Sakurai, T. Torii, N. Ohsako, and H. Sugiura, *A method for finding clusters of zeros of analytic function*, Special Issues of Zeitschrift für Angewandte Mathematik und Mechanik (ZAMM). Issue 1: Numerical Analysis, Scientific Computing, Computer Science, 1996, Proceedings of the International Congress on Industrial and Applied Mathematics (ICIAM/GAMM 95), Hamburg, July 3–7, 1995, pp. 515–516.
112. A. Sidi, *A new variable transformation for numerical integration*, Numerical Integration IV (H. Brass and G. Hämmerlin, eds.), International Series of Numerical Mathematics, vol. 112, Birkhäuser Verlag, 1993, pp. 359–373.
113. I. H. Sloan and S. Joe, *Lattice methods for multiple integration*, Clarendon, Oxford, 1994.
114. V. I. Smirnov and N. A. Lebedev, *Functions of a complex variable: Constructive theory*, Iliffe Books, 1968.
115. G. W. Stewart, *Perturbation theory for the generalized eigenvalue problem*, Recent Advances in Numerical Analysis (C. De Boor and G. H. Golub, eds.), Academic Press, 1978, pp. 193–206.
116. J. F. Traub, G. W. Wasilkowski, and H. Woźniakowski, *Information-based complexity*, Academic Press, 1988.
117. E. E. Tyrtyshnikov, *How bad are Hankel matrices?*, Numer. Math. **67** (1994), 261–269.
118. W. Van Assche, *Orthogonal polynomials in the complex plane and on the real line*, Special Functions, q-Series and Related Topics (M. E. H. Ismail, D. R. Masson, and M. Rahman, eds.), Fields Institute Communications, vol. 14, American Mathematical Society, Providence, Rhode Island, 1997, pp. 211–245.

119. M. Van Barel and A. Bultheel, *A new approach to the rational interpolation problem*, J. Comput. Appl. Math. **32** (1990), 281–289.

120. P. Van Hentenryck, D. McAllester, and D. Kapur, *Solving polynomial systems using a branch and prune approach*, SIAM J. Numer. Anal. **34** (1997), no. 2, 797–827.

121. J. Verschelde and R. Cools, *Polynomial homotopy continuation, A portable Ada software package*, The Ada-Belgium Newsletter **4** (1996), 59–83.

122. C. Von Westenholz, *Differential forms in mathematical physics*, Studies in Mathematics and its Applications, vol. 3, North-Holland, Amsterdam, 1978.

123. G. N. Watson, *A treatise on the theory of Bessel functions*, second ed., Cambridge University Press, 1966.

124. J. H. Wilkinson, *The evaluation of the zeros of ill-conditioned polynomials. Part I*, Numer. Math. **1** (1959), 150–166.

125. J.-Cl. Yakoubsohn, *Approximating the zeros of analytic functions by the exclusion algorithm*, Numerical Algorithms **6** (1994), 63–88.

126. X. Ying, *A reliable root solver for automatic computation with application to stress analysis of a composite plane wedge*, D. Sc. dissertation, Washington University in St. Louis, 1986.

127. X. Ying and I. N. Katz, *A reliable argument principle algorithm to find the number of zeros of an analytic function in a bounded domain*, Numer. Math. **53** (1988), 143–163.

128. _____, *A simple reliable solver for all the roots of a nonlinear function in a given domain*, Computing **41** (1989), no. 4, 317–333.

Druck: Strauss Offsetdruck, Mörlenbach
Verarbeitung: Schäffer, Grünstadt